Springer Series on
ATOMIC, OPTICAL, AND PLASMA PHYSICS 52

Springer Series on
ATOMIC, OPTICAL, AND PLASMA PHYSICS

The Springer Series on Atomic, Optical, and Plasma Physics covers in a comprehensive manner theory and experiment in the entire field of atoms and molecules and their interaction with electromagnetic radiation. Books in the series provide a rich source of new ideas and techniques with wide applications in fields such as chemistry, materials science, astrophysics, surface science, plasma technology, advanced optics, aeronomy, and engineering. Laser physics is a particular connecting theme that has provided much of the continuing impetus for new developments in the field. The purpose of the series is to cover the gap between standard undergraduate textbooks and the research literature with emphasis on the fundamental ideas, methods, techniques, and results in the field.

R. Boudet

Relativistic Transitions in the Hydrogenic Atoms

Elementary Theory

 Springer

Roger Boudet
Université de Provence
7 Av. de Servian
34290 Bassan
France
E-mail: boudet@cmi.univ-mrs.fr

Springer Series on Atomic, Optical, and Plasma Physics   ISSN 1615-5653

ISBN 978-3-540-85549-1   e-ISBN 978-3-540-85550-7

Library of Congress Control Number: 2008934401

Typesetting and production: SPi Publisher Services
Cover concept: eStudio Calmar Steinen
Cover design: WMX Design GmbH, Heidelberg

SPIN 12445820   57/3180/SPi
Printed on acid-free paper

9 8 7 6 5 4 3 2 1

springer.com

Preface

The aim of this volume is twofold. First, it is an attempt to simplify and clarify the relativistic theory of the hydrogen-like atoms. For this purpose we have used the mathematical formalism, introduced in the Dirac theory of the electron by David Hestenes, based on the use of the real Clifford algebra $Cl(M)$ associated with the Minkwoski space–time M, that is, the euclidean R^4 space of signature (1,3). This algebra may be considered as the extension to this space of the theory of the Hamilton quaternions (which occupies an important place in the resolution of the Dirac equation for the central potential problem).

The clarity comes from the real form given by D. Hestenes to the electron wave function that replaces, in a strict equivalence, the Dirac spinor. This form is directly inscribed in the frame of the geometry of the Minkwoski space in which the experiments are necessarily placed. The simplicity derives from the unification of the language used to describe the mathematical objects of the theory and the data of the experiments.

The mathematics concerning the definition and the use of the algebra $Cl(M)$ are not very complicated. Anyone who knows what a vector space is will be able to understand the geometrical implications of this algebra. The lecture will be perhaps more difficult for the readers already acquainted with the complex formalism of the matrices and spinors, to the extent that the new language will appear different from the one that they have used. But the correspondence between the two formalisms is ensured in the text at each stage of the theory.

The second aim concerns a presentation of the theory of one-electron atoms starting from its relativisitic foundation, the Dirac equation. The nonrelativistic Pauli and Schrödinger theories are introduced as approximations of this equation. One of the major purpose, about these approximations, has been to display, on the one side, the enough good concordance between the Dirac and the Pauli–Schrödinger theories for the bound states of the electron furthermore, but to a weaker extent, for the states of the continuum close to the freedom energy, and, on the other side, the considerable discordances for the high values of the continuum. A special attention has been drawn to the verification of the numerical relativistic results by the comparison with those obtained by means of the nonrelativistic approximations, when the comparison is acceptable, and also to the recourse to different mathematical methods for the resolution of a same problem.

Bassan,   *Roger Boudet*
August 2008

Contents

Part II Fields Created by the Dirac Transition Currents Between Two States

Part V Interaction with a Magnetic Field

Part VI Addendum

Part VII Appendices

1

Introduction

The present volume is devoted to the transitions in hydrogen-like atoms, also called hydrogenic atoms. One means by hydrogenic atom, an atom considered as owning one electron as the hydrogen atom. It is the case for an atom whose all the electrons, except one, are not considered, either because they have been thrown out or because their action is neglected.

Corrective terms taking into account this action, or the size of the nucleus, may be used. But they are obtained by means of approximative instead of exact calculations, and they will not be taken into consideration in our elementary presentation.

So the problem, in its relativistic approach, is the first one of the resolution of the Dirac equation for a central potential of the form eZ/r, where $e > 0$ is the charge of a proton (with $-e$ as the charge of the electron) and Z is the number of protons in the nucleus of the atom. The question of the transitions between two states of the electron is solved by taking into account the two solutions of the Dirac equation corresponding to these states, by the construction of the probability current of transition between these two states and at least by the determination of the electromagnetic field at large distance associated with this current. The processes of the transition are also to be taken into account: spontaneous emission in the absence of all external field, stimulated transition in the presence of an external plane wave, and multiplication of the transitions in the case where a magnetic field separates into several levels of energy, the level common to the states corresponding to a same level in the absence of a magnetic field (Zeeman effect).

Part I

The Solutions of the Dirac Equation
in Hydrogenic Atoms

2

The Electromagnetic Fields Created
by Time-Sinusoidal Current

Abstract. This chapter is a recall of the properties of the long-range part of the electromagnetic fields created by time-periodic currents, as they may be observed in particular in the Zeeman effect. The aim of this part is also to place the vector frame of these observations, that is, one of the spherical coordinates, which is in the center of the presentation in the real formalism of the relativistic central potential problem. This frame is the one in which are expressed the Dirac probability current, associated with a state and with the transition between two states. But it is to notice that, as a specificity of the real formalism, the form given by Hestenes to the wave function of the electron, strictly equivalent to the Dirac spinor, may be presented, in the case of central potential, as a combination of the vectors of this frame.

2.1 Properties of the Electromagnetic Field Emitted
by an Electron Bound in an Atom

The observation of the electromagnetic fields emitted by electrons bound in an atom, achieved when a magnetic field is present (Zeeman effect), shows that the field owns the following particularities:

1. The field is time-sinusoidal and polarized.
2. If the observation is orthogonal to the direction of the magnetic field, the polarization appears as being linear along this direction.
3. If the observation is parallel to this direction, the polarization appears as being circular and in a plane orthogonal to this direction.

Such data of the observations allow one to precise the general form of the electric currents, which are the source of the field.

The extension of these particularities to the transitions where no magnetic field is present, that is, spontaneous or stimulated emissions processes, is not directly observable. But it is confirmed not only by other experimental data, but also by the fact that the theoritical construction of the transition currents is deduced from the Darwin solutions of the Dirac equation, and that these solutions give exactly (if one excepts the small variation called the Lamb shift) the values of the levels of energy of an electron bound in an atom.

2.2 The Field at Large Distance of a Time-Periodic Current

The calculation of the field that is used here is based on the pure laws of Maxwell, without quantization. Indeed, using Quantum Field Theory is not a necessity in the domain studied here. It leads exactly to the same results (see [12]), with sometimes longer calculations.

We consider only the long-range part of the field by applying the following theorem [41]. If the source of the field is negligible outside a small neighbourhood of the origin O, the long-range part of the field is deduced from the integral formula of the retarded potential in such a way that

$$\mathbf{E}(x^0, \mathbf{r}) = -q \frac{\partial}{\partial x^0} \int \frac{\mathbf{j}^\perp(x^0 - R, \mathbf{r}\,')}{R} \, \mathrm{d}\tau', \tag{2.1}$$

$$\mathbf{H}(x^0, \mathbf{r}) = -q \frac{\partial}{\partial x^0} \int \frac{\mathbf{n} \times \mathbf{j}^\perp(x^0 - R, \mathbf{r}\,')}{R} \, \mathrm{d}\tau', \tag{2.2}$$

where the coordinates x^μ are in the form $(x^0 = ct, \mathbf{r})$ and q is the charge of the source.

The vector $\mathbf{j}^\perp$ is the component of the spatial part $\mathbf{j} = (j^1, j^2, j^3)$ of the space–time vector j^μ, orthogonal to the vector $\mathbf{n} = \mathbf{R}/R$, where $\mathbf{R} = \mathbf{r} - \mathbf{r}'$, $R = |\mathbf{R}|$. Note that the time component j^0 of the current does not intervene.

In the theory of the electron, the vector j^μ has the meaning of a current of probability of the presence of the electron and $q = -e$ is the charge.

We can notice furthermore that if j^μ in time-independent, the long-range part of the field is null. As it is the case of the Dirac probability current j^μ associated with the state of a bound electron, this explains the reason why no electromagnetic field may be observed outside a passage from a state to one another.

If the field is time-sinusoidal, the source current is of the form

$$q \, \mathbf{j}(x^0, \mathbf{r}) = q \left[\cos \omega x^0 \, \mathbf{j}_1(\mathbf{r}) + \sin \omega x^0 \, \mathbf{j}_2(\mathbf{r})\right], \tag{2.3}$$

where the vectors $\mathbf{j}_k$ are to be precised.

At large distance r from the origin O, we may replace $\mathbf{r} - \mathbf{r}'$ by $\mathbf{r} = r\mathbf{n}$ in (2.1) and write

$$\mathbf{E}(x^0, \mathbf{r}) = -\frac{q}{r} \frac{\partial}{\partial x^0} \int \mathbf{j}^\perp(x^0 - r, \mathbf{r}') \, \mathrm{d}\tau' \tag{2.4}$$

and so we can write

$$E(x^0, \boldsymbol{r}) = q\frac{\omega}{r}[\sin\omega(x^0 - r)\, \boldsymbol{U}_1^{\perp} - \cos\omega(x^0 - r)\, \boldsymbol{U}_2^{\perp}], \qquad (2.5)$$

where

$$\boldsymbol{U}_k^{\perp} = \int \boldsymbol{j}_k^{\perp}(\boldsymbol{r}')\, d\tau', \quad k = 1, 2. \qquad (2.6)$$

2.3 Source Currents of Time-Sinusoidal Polarized Field

Let $(\boldsymbol{e}_1, \boldsymbol{e}_2, \boldsymbol{e}_3)$ be an orthogonal frame of the three-space of the laboratory galilean frame. The most convenient coordinates system for defining the current is the (r, θ, φ) spherical coordinate system, in which the vector $\boldsymbol{e}_3$ defines a privileged direction, the one of the magnetic field in the case of the presence of this field,

$$\begin{aligned}\boldsymbol{u} &= \cos\varphi\, \boldsymbol{e}_1 + \sin\varphi\, \boldsymbol{e}_2, \quad \boldsymbol{v} = -\sin\varphi\, \boldsymbol{e}_1 + \cos\varphi\, \boldsymbol{e}_2, \\ \boldsymbol{n} &= \cos\theta\, \boldsymbol{e}_3 + \sin\theta\, \boldsymbol{u}, \quad \boldsymbol{w} = -\sin\theta\, \boldsymbol{e}_3 + \cos\theta\, \boldsymbol{u}, \quad \boldsymbol{r} = r\boldsymbol{n}. \end{aligned} \qquad (2.7)$$

For taking into account the polarizations, the components $\boldsymbol{j}_1, \boldsymbol{j}_2$ of the current may be then defined in the following way:

$$\boldsymbol{j}_1 = \cos\epsilon\varphi\, \boldsymbol{j}_I + \sin\epsilon\varphi\, \boldsymbol{j}_{II}, \quad \boldsymbol{j}_2 = -\sin\epsilon\varphi\, \boldsymbol{j}_I + \cos\epsilon\varphi\, \boldsymbol{j}_{II}, \qquad (2.8)$$

where

$$\boldsymbol{j}_I = b(r, \theta)\, \boldsymbol{v}, \quad \boldsymbol{j}_{II} = a(r, \theta)\, \boldsymbol{u} + c(r, \theta)\, \boldsymbol{e}_3, \qquad (2.9)$$

and where ϵ may be taken equal to 0 or ± 1. We consider the vector

$$\boldsymbol{U} = \cos\omega x^0\, \boldsymbol{U}_1 + \sin\omega x^0\, \boldsymbol{U}_2, \quad \boldsymbol{U}_k = \int \boldsymbol{j}_k(\boldsymbol{r})\, d\tau. \qquad (2.10)$$

2.3.1 Linear Polarization: $\epsilon = 0$

In this case we have $\boldsymbol{j}_1 = \boldsymbol{j}_I$ and $\boldsymbol{j}_2 = \boldsymbol{j}_{II}$. The relations $d\tau = (r\sin\theta d\varphi)(rd\theta)dr$ and $\int_0^{2\pi} \boldsymbol{u}\, d\varphi = 0 = \int_0^{2\pi} \boldsymbol{v}\, d\varphi$ give

$$\boldsymbol{U}_1 = 0,\ \boldsymbol{U}_2 = C\boldsymbol{e}_3,\ C = 2\pi\int_0^{\infty}\int_0^{\pi} c(r, \theta)r^2\sin\theta\, drd\theta \quad \boldsymbol{U} = \sin\omega x^0\, C\boldsymbol{e}_3. \qquad (2.11)$$

2.3.2 Circular Polarizations: $\epsilon = \pm 1$

In this case we deduce immediately

$$\begin{aligned}\boldsymbol{j}_1^{\pm} &= (\pm a - b)\cos\varphi\sin\varphi\, \boldsymbol{e}_1 + (b\cos^2\varphi \pm a\sin^2\varphi)\, \boldsymbol{e}_2 \pm c\sin\varphi\, \boldsymbol{e}_3, \\ \boldsymbol{j}_2^{\pm} &= (\pm b\sin^2\varphi + a\cos^2\varphi)\, \boldsymbol{e}_1 + (a \mp b)\cos\varphi\sin\varphi\, \boldsymbol{e}_2 + c\cos\varphi\, \boldsymbol{e}_3 \end{aligned}$$

and after integration

$$U_1^\pm = (J_I \pm J_{II})\, e_2, \quad U_2^\pm = (\pm J_I + J_{II})\, e_1,$$
$$U^+ = A(\sin \omega x^0\, e_1 + \cos \omega x^0\, e_2), \quad A = J_{II} + J_I,$$
$$U^- = B(\sin \omega x^0\, e_1 - \cos \omega x^0\, e_2), \quad B = J_{II} - J_I, \tag{2.12}$$
$$J_I = \pi \int_0^\infty \int_0^\pi b(r, \theta) r^2 \sin \theta\, dr d\theta, \quad J_{II} = \pi \int_0^\infty \int_0^\pi a(r, \theta) r^2 \sin \theta\, dr d\theta.$$

2.4 Flux of the Poynting Vector Through a Sphere of Large Radius

Let us consider the flux F, per unit of time, through a sphere S of large radius, of the Poynting vector of the field, created by the transition current between two states, of an electron bound in an atom. If we consider the energy E released at each transition, the ratio F/E gives the number of transitions per second.

If no external field is present, the transition is called spontaneous emission. The number of these transitions may be experimentally observed, and, for comparison, the theoretical calculation presents an interest (see Chap. 7).

We consider that F is averaged on a period $T = 2\pi w$ of the source current and denoted by

$$\langle X \rangle = \frac{1}{T} \int_0^T X\, dx^0,$$

the average of X.

Because E and $H = n \times E$ are orthogonal to n, we can write for a sphere S of center 0 of radius R

$$F = \frac{c}{4\pi} \int_{S_0} \langle n \cdot (E \times H) \rangle R^2 d\sigma,$$

then

$$F = \frac{c}{4\pi} \int_{S_0} \langle E^2 \rangle R^2 d\sigma, \tag{2.13}$$

where S_0 is the sphere unity. Now

$$\langle \cos^2 \omega(x^0 - R) \rangle = \langle \sin^2 \omega(x^0 - R) \rangle = \frac{1}{2},$$
$$\langle \cos 2\omega(x^0 - R) \sin \omega(x^0 - R) \rangle = 0.$$

In other respect, let (θ_0, φ_0) be the system of spherical coordinates of S_0, such that the axis of the poles is colinear with one of the vectors U_k.

We can write $(U_k^\perp)^2 = U_k^2 \sin^2 \theta_0$ and

$$\int_{S_0} (U_k^\perp)^2 \mathrm{d}\sigma = \int_0^{2\pi} \int_0^\pi [U_k^2 \sin^2 \theta_0] \sin \theta_0 \, \mathrm{d}\varphi \mathrm{d}\theta_0 = \frac{8\pi}{3} U_k^2, \qquad (2.14)$$

and taking into account the presence of $1/R^2$ in $\boldsymbol{E}^2$, we can replace in all the cases of polarization (2.13) by the equation

$$F = \frac{c\omega^2 e^2}{3}(U_1^2 + U_2^2). \qquad (2.15)$$

2.5 Units

The only constants we use are the three fundamental constants (revised in 1989 by B.N. Taylor):

1. The light speed $c = 2.99\ 792\ 458 \times 10^{10}\ \mathrm{cm\,s^{-1}}$.
2. The electron charge magnitude $e = 4.803\ 206 \times 10^{-10}$ (e.s.u.).
3. The reduced Planck constant $\hbar = h/2\pi = 1.054\ 572 \times 10^{-27}\ \mathrm{erg\,s}$. In addition we use
4. the electron mass $m = 9.109\ 389 \times 10^{-28}\ \mathrm{g}$. All the other constants used will be derived from these four ones, in particular,
5. the fine structure constant

$$\alpha = \frac{e^2}{\hbar c} = \frac{1}{137.035\ 989} \qquad (e \text{ in e.s.u.}) \qquad (2.16)$$

and as unit of length:
6. the "radius of first Bohr orbit"

$$a = \hbar^2/(me^2) = \hbar/(mc\alpha) = 5.291\ 772 \times 10^{-9}\ \mathrm{cm}. \qquad (2.17)$$

Note

In other respects, one introduces in the expression of the electromagnetic potentials the factor $1/(4\pi\epsilon_0)$ (the presence of 4π is due to the writing $4\pi j^\mu$ instead of j^μ in the current term of the Maxwell equations), where ϵ_0 is the permittivity of free space, and e is expressed in e.m.u:

$$\epsilon_0 = 8.854187 \times 10^{-12}\ \mathrm{F\ m^{-1}}, \quad e = 1.602\ 1777 \times 10^{-19}\ (\text{e.m.u.})$$

That gives (with c expressed in metres) the same value of α with the expression

$$\alpha = \frac{e^2}{4\pi\epsilon_0 \hbar c}, \qquad (e \text{ in e.m.u.}) \qquad (2.18)$$

For simplicity and to be in agreement with the largest part of the reference articles and treatises mentioned here, we use the former expressions of the potentials and the constant α, in preference to these last ones.

3

The Dirac Equation of the Electron in the Real Formalism

Abstract. This chapter is a recall of the algebraic tool used in the real formalism and the passage from the Dirac equation of the electron in the complex spinor formalism to the real form given by Hestenes to this equation. It is completed by the first paragraph of the Appendix of Part I, which allows a step-by-step traduction of the complex formalism to the real one and vice versa.

3.1 Algebraic Preliminaries: A Choice of Formalism

The usual presentation of the Dirac theory of the electron uses, on the one side, the σ_k and γ_μ Pauli and Dirac *matrices*, and on the other side, the Pauli and Dirac *spinors*. The mathematical language of this presentation is faraway from one of the experiments described in Chap. 2.

In contrary, the real formalism we present further uses the same mathematical objects as the ones employed in this section. In particular, in the case of the solutions of the Dirac equation for the central potential problem, the wave function of the electron may be directly expressed by means of the same vectors as in (2.7), that is, those of the frame of the spherical coordinates in $E^3 = R^{3,0}$.

The Pauli and Dirac matrices may be related to the geometry of the E^3 and the Minkowski $M = R^{1,3}$ spaces, respectively. Despite their *complex* and complicated forms, they obey relations similar to the g_{ij} ones verified by the orthonormal frames of these *real* spaces.

Futhermore, it is well known that these matrices allow one to construct spaces isomorphic to the so-called Clifford algebras $Cl(E^3) = Cl(3,0)$ (or ring of the biquaternions) and $Cl(M) = Cl(1,3)$, associated with E^3 and M, respectively (see later).

These isomorphisms become identities as soon as the sets $\{\sigma_k\}$ and $\{\gamma_\mu\}$ are identified, as was implicitly made by some authors, to an orthonormal frame of E^3 and M, respectively.

We recall (see Appendix D) that a Clifford algebra $Cl(p, n - p)$ is a *real* associative algebra, acting on the vectors of an euclidean space $R^{p,n-p}$, which associates the signature of this space with the elements of the Grassmann (or exterior) algebra $\wedge R^n$ in such a way that not only the elements of this algebra are geometrical objects, but furthermore they may be used for defining geometrical operations (in particular the isometries) upon these objects.

But the Pauli and the Dirac *spinors* remain *abstract objects*, implying the use of a geometrically undefined "number" $\sqrt{-1}$.

In a fundamental article [35], Professor David Hestenes has introduced what he calls the Algebra of Space–Time (STA) in the Dirac theory of electron. STA is the Clifford algebra $Cl(M)$ considered for its applications in Quantum Mechanics.

But, and that was entirely new, D. Hestenes replaces, *in a strict equivalence*, the Dirac spinor Ψ by a pure real geometrical object ψ that we call for its applications in Quantum Mechanics a *Hestenes spinor*: an element of the even sub-algebra $Cl^+(M)$ of $Cl(M)$. As a consequence, *the set without any proper structure of the Dirac spinors is replaced by a ring*, and also the matrices and the spinors *are unified in a same structure*, implying only geometrical elements of the Minkowksi *real* space M.

The fact that the Pauli and the Dirac *spinors* (we repeat, *spinors* not *matrices*) are nothing else but a decomposition in two and four "complex numbers" of an element of $Cl^+(E^3)$ and $Cl^+(M)$ is recalled in Appendix A.1 and A.2.

The ring $Cl^+(M)$ may be identified with the ring $Cl(E^3)$ of the biquaternions, by the fact that, if $\{e_\mu\}$ is an orthonormal frame of M, the set of the bivectors of M $\{e_k = e_k \wedge e_0\}$, $k = 1, 2, 3$, generates a E^3 space, used, for example, in the writing $F = \boldsymbol{E} + i\boldsymbol{H}$ of the electromagnetic field $F \in \wedge^2 M$.

The subalgebra $Cl^+(E^3)$ of $Cl(E^3)$ is the *field* of the Hamilton quaternions. We see that this field plays an important role in the theory of the Dirac electron in a central potential.

Let us consider the rings $Cl(3, 0) \simeq Cl^+(1, 3)$ and $Cl(1, 3)$ as algebraic continuations of the field $Cl^+(3, 0)$. Given the fact that the field of the Hamilton quaternions $Cl^+(3, 0)$ is *the only field* that may be associated with a space R^n for $n > 2$, one sees that *the signature (1,3) of the R^4 space and time appears as algebraically priviledged.*

More surprising is the fact *that the stucture of all the objects, quarks, and leptons, defined by means of the Dirac spinor–in the form given by D. Hestenes–which fill this 4-space, also appears as algebraically priviledged.*

It is to emphasize the simplification and the geometrical clarity that the use of the real Hestenes formalism brings not only to Quantum Mechanics but also to the other domains of physics with the name of Geometric Algebra [22, 37].

All the observed phenomenas of the special relativity are necessarily placed in the scope of the pure geometry of space–time, and all the mathematical objects whose aim is to interpret these phenomenas are placed by the real formalism directly in this scope.

In particular, for the case of the central potential, all these objects and in particular the Dirac spinor will be expressed, after a passage from $Cl^+(1,3)$ to $Cl(3,0)$, by means of vectors of $R^{3,0}$.

However, the correspondence between this real formalism and the complex one of the Pauli and Dirac matrices and spinors will be ensured here at each step of the calculation, in such a way that a reader acquainted with one of these two formalisms will be able to use the other.

Let us mention that biquaternionic solutions of the central potential problem had been already achieved by Sommerfeld (see [54]), but in the frame of the complex formalism, the Dirac spinor being expressed by means of the Dirac matrices γ^μ. Nevertheless, though, in particular, some ambiguities related to the role of the "imaginary number" $\sqrt{-1}$ in the complex formalism had been removed in this work, the use of the pure real formalism of Hestenes brings noticeable simplifications and above all the entire geometrical clarification of the theory of the electron.

3.1.1 Quaternions and Biquaternions

1. The main properties of all Clifford algebra $Cl(p, n-p) = Cl(E)$ associated with a euclidean space $E = R^{p,n-p}$ are recalled in Appendix D.

Let us mention that only $Cl(E)$ is an associative algebra of dimension 2^n acting on the vectors of E and related to the following:

- The euclidean structure of E. If $a_1 a_2 \cdots a_p$ denotes the Clifford product of vectors $a_k \in E$, the scalar product $a \cdot b$ of two vectors verifies

$$a.b = \frac{1}{2}(ab + ba). \qquad (3.1)$$

- The Grassmann (or exterior algebra) $\wedge R^n$. If the vectors a_k are orthogonal, the Clifford product $a_1 a_2 \cdots a_p$ is equal to their Grassmann product $a_1 \wedge a_2 \cdots \wedge a_p$ (see Appendix D).

In particular, if two vectors a, b are orthogonal, one obtains the important rule

$$a.b = 0 \implies ab = a \wedge b = -b \wedge a = -ba. \qquad (3.2)$$

These algebras, in particular, $Cl(3,0)$ (or ring of the biquaternions) and $Cl(1,3)$, and their even subalgebras $Cl^+(3,0)$ (or field of the quaternions), $Cl^+(1,3) \simeq Cl(3,0)$ (see later), may be used directly for the study of Quantum Mechanics.

What follows in the present paragraph concerns readers already acquainted with the complex formalism of the Pauli and Dirac matrices, in the purpose to give an indication on the links of these matrices with these algebras.

For the lecture of the present work, the readers can refer directly to Sect. 3.1.2 below, then to the real formalism of the Clifford algebras and avoid all that concerns the correspondence between this formalism and the one of the complex matrices and spinors. However, the knowlewge of this correspondence is recommended.

2. There exists a well known construction of the real Clifford algebra $Cl(E^3) = Cl(3,0)$, or ring of the biquaternions, associated with the space $E^3 = R^{3,0}$ by means of the Pauli matrices σ_k (see [54], IV.5.1).

One considers a *real* space of dimension $1 + 3 + 3 + 1 = 8$ but whose a frame is composed by the following entities:

$$I$$

$$\sigma_1, \sigma_2, \sigma_3$$

$$\underline{i}\sigma_1 = \sigma_2\sigma_3, \underline{i}\sigma_2 = \sigma_3\sigma_1, \underline{i}\sigma_3 = \sigma_1\sigma_2$$

$$\underline{i} = \sigma_1\sigma_2\sigma_3,$$

where I is the unit (2×2) matrix.

We say that this space may be considered as real because all element of this space is a linear combination of the aforementioned entities with *real* coefficients. This construction is based on the following joint properties.

Let $\{e_1, e_2, e_3\}$ be an orthonormal frame of E^3. One may notice that the Pauli matrices σ_k obey the same rule as the vectors e_k in $Cl(E^3)$

$$\frac{1}{2}(\sigma_i\sigma_j + \sigma_j\sigma_i) = \delta_{ij}I, \quad e_i \cdot e_j = \frac{1}{2}(e_ie_j + e_je_i) = \delta_{ij},$$

and an identification of these matrices to these vectors (the unit matrix I being identified to 1) explains the possibility of the construction of $Cl(E^3)$ by means of the matrices σ_k.

The subspaces generated by I, σ_k, $\underline{i}\sigma_k$, $\underline{i}$ correspond to the subspaces of $\wedge R^3$, scalar, vector, bivector, and pseudo-scalar spaces of R^3.

The space generated by I and $\underline{i}\sigma_k$ (scalars plus bivectors), whose dimension is $1 + 3 = 4$, corresponds to the even subalgebra $Cl^+(E^3)$ of $Cl(E^3)$, that is the *field H* of the Hamilton quaternions.

3. A *real* space of dimension $1 + 4 + 6 + 4 + 1 = 16$ generated in the same way by means of the Dirac matrices γ_μ and the similitude of the properties of these matrices with the vectors e_μ of an orthonormal frame of the Minkowski space $M = R^{1,3}$ explain the possibility of the construction of $Cl(M)$ by means of the matrices γ_μ (see [6], (2.5)).

The subspaces of this algebra correspond to the subspaces of $\wedge R^4$, scalar, vector, bivector, pseudo-vector, and pseudo-scalar spaces of R^4.

The sum of the scalar, bivector, pseudo-scalar spaces whose dimension is $1 + 6 + 1 = 8$ is the even subalgebra $Cl^+(M)$ of $Cl(M)$. It may be identified to $Cl(E^3)$ by using the relation $\sigma_k = \gamma_k\gamma_0$ (see Sect. 3.1.3).

The identifications

$$\sigma_k \Longleftrightarrow e_3, \quad \gamma_\mu \Longleftrightarrow e_\mu$$

allow the identification of the spaces constructed by means of σ_k and γ_μ to the entirely real spaces $Cl(E^3)$ and $Cl(M)$.

The links of the Pauli and the Dirac *spinors* with $Cl^+(E^3)$ and $Cl^+(M)$ are less evident. These spinors are eachone the decompostion into two and four complex numbers of an element of $Cl^+(E^3)$ and $Cl^+(M)$. These numbers are written in the form $a + ib$ in the complex formalism, but the "imaginary number" $i = \sqrt{-1}$ is in fact $\underline{i}\sigma_3 = \gamma_2\gamma_1$ (see Sect. 3.1.3) which becomes real by the above identifications. The presence of $\gamma_2\gamma_1$, that is, the bivector of M, $e_2e_1 = e_2 \wedge e_1$ (whose square in $Cl(M)$ is equal to -1) in place of the imaginary number i is closely related to the existency of the spin of the electron.

The conversion of the spinors in elements of $Cl^+(E^3)$ and $Cl^+(M)$ has been established in [35], and is recalled in detail in Appendix A, following the method introduced in [28] and recalled in [44].

3.1.2 The Hamilton Quaternion and the Pauli Spinor

Following (with a change of sign upon i, j, k) the definition given in 1849 by Hamilton, a quaternion q verifies the relations

$$q = d + ia + jb + kc, \quad (a, b, c, d \in R), \tag{3.3}$$

$$i^2 = j^2 = k^2 = -1; \quad i = kj = -jk, \; j = ik = -ki, k = ji = -ij, \tag{3.4}$$

with the following geometrical interpretation

$$i = e_2e_3 = e_2 \wedge e_3, \quad j = e_3e_1 = e_3 \wedge e_1, \quad k = e_1e_2 = e_1 \wedge e_2, \tag{3.5}$$

where $\boldsymbol{a} \wedge \boldsymbol{b}$ means the *Grassmann* (not the vector) product of $\boldsymbol{a}, \boldsymbol{b}$ (see Appendix D). So a quaternion is a *real* object, the sum of a scalar and a bivector of E^3.

Using the Clifford product in $Cl(E^3)$, denoting

$$\underline{i} = e_1 \wedge e_2 \wedge e_3 = e_1e_2e_3 \tag{3.6}$$

one may write because, for example, $e_3^2 = 1$, $k = e_1 \wedge e_2 = e_1e_2 = \underline{i}e_3$,

$$i = \underline{i}e_1, \quad j = \underline{i}e_2, \quad k = \underline{i}e_3. \tag{3.7}$$

As a consequence, a quaternion may be written in the form $q = d + \underline{i}\boldsymbol{a}$, $\boldsymbol{a} \in E^3$.

It is easy to deduce from the relations $\underline{i}e_k = e_k\underline{i}$ that

$$\underline{i}\boldsymbol{a} = \boldsymbol{a}\underline{i}, \quad \underline{i}^2 = -1 \tag{3.8}$$

We can notice that different *real* objects, $i, j, k, \underline{i}$, may own the same important property: *their square is equal to* -1.

The relation (see Appendix D)

$$\boldsymbol{ab} = \boldsymbol{a} \cdot \boldsymbol{b} + \boldsymbol{a} \wedge \boldsymbol{b}, \quad \boldsymbol{a}, \boldsymbol{b} \in E^3$$

shows that all Clifford product $\boldsymbol{ab}$ of two vectors of E^3 is a quaternion.

Let us denote by $\boldsymbol{a} \times \boldsymbol{b}$ the *vector* product in E^3, as for example $\boldsymbol{e}_3 = \boldsymbol{e}_1 \times \boldsymbol{e}_2$, from which we deduce ((3.5) and (3.7)) that $\boldsymbol{e}_1 \wedge \boldsymbol{e}_2 = \underline{i}(\boldsymbol{e}_1 \times \boldsymbol{e}_2)$ and that we can write

$$\boldsymbol{a} \wedge \boldsymbol{b} = \underline{i}(\boldsymbol{a} \times \boldsymbol{b}). \tag{3.9}$$

So q takes the form $q = \boldsymbol{a} \cdot \boldsymbol{b} + \underline{i}(\boldsymbol{a} \times \boldsymbol{b})$, a writing similar to a formula often used in the theory of the Pauli matrices.

In what follows the bivector $k = \underline{i}e_3$ is going to take an important place, and one of the reason lies in the fact that the $\boldsymbol{e}_3$-axe owns an important place in the theory of the Dirac electron, in particular, when it is applied to the study of hydrogenic atoms. Another reason, the fact that $\underline{i}e_3$ defines also a bivector of spacetime (and we repeat that the presence of this bivector is related to the existency of the spin of the electron), will be evoked later.

The identification of a Pauli *spinor* $\xi \in C^2$ to a Hamilton quaternion q (3.3) may be achieved (see Appendix A.1) by the following rule:

$$\begin{aligned}
\xi = (u_1', u_2') &\iff q = u_1 - ju_2, \quad \sqrt{-1} \iff k, \\
u_1' &\iff u_1 = d + kc, \quad u_2' \iff u_2 = -b + ka, \\
-jk &= i = \underline{i}e_1, \quad j = \underline{i}e_2, \quad k = \underline{i}e_3.
\end{aligned} \tag{3.10}$$

So the *ring* of the $\underline{i}\sigma_k$ Pauli *matrices* and the *set* without any proper structure of the Pauli *spinors* may be unified in a single real field, the *only field* one can associate with a real space R^n for $n > 2$.

3.1.3 The Hestenes Spinor and the Dirac Spinor

1. The field $H = Cl^+(3, 0)$ of the Hamilton quaternions may be extended to the ring of the Clifford biquaternions $Cl(3, 0)$, whose all element Q may be written as

$$Q = q_1 + \underline{i}q_2, \quad q_\alpha = d_\alpha + \underline{i}a_\alpha. \tag{3.11}$$

Let $\{e_\mu\}$ be an orthonormal frame of the Minkowski space $M = R^{1,3}$.

The bivectors $e_k \wedge e_0$ of M may be associated with the vectors $\boldsymbol{e}_k$

$$\boldsymbol{e}_1 = e_1 \wedge e_0 = e_1 e_0, \quad \boldsymbol{e}_2 = e_2 \wedge e_0 = e_2 e_0, \quad \boldsymbol{e}_3 = e_3 \wedge e_0 = e_3 e_0. \tag{3.12}$$

They generate a space E^3, associated with e_0, which allows the identification of $Cl^+(1, 3) = Cl^+(M)$ to the ring $Cl(3, 0) = Cl(E^3)$ of the Clifford biquaternions, with the rule

$$a \in M, \quad a.e_0 = 0 \implies \boldsymbol{a} = a \wedge e_0 = ae_0 \in Cl^+(M), \quad \boldsymbol{a} \in E^3. \tag{3.13}$$

We recall that such a rule allows one the well known writing $F = \boldsymbol{E} + \underline{i}\boldsymbol{H}$ of the electromagnetic field $F \in \wedge^2 M$.

2. We use the following operation of reversion:

$$X \in Cl(M) \;\rightarrow\; \tilde{X} \in Cl(M), \text{ so that } (XY)\tilde{} = \tilde{Y}\tilde{X}, \qquad (3.14)$$

with $\tilde{\lambda} = \lambda$ and $\tilde{a} = a$ if $\lambda \in R$, $a \in M$.

An important remark is the following. One can write, both in $Cl(E^3)$ and $Cl(M)$,

$$\underline{i} = e_1 e_2 e_3 = e_0 e_1 e_2 e_3 \;\Rightarrow\; \underline{i}^2 = -1,$$
$$(\underline{i}e_3)^2 = (e_1 e_2)^2 = (e_2 e_1)^2 = -1. \qquad (3.15)$$

It is also important to notice that

$$\underline{i}\boldsymbol{a} = \boldsymbol{a}\underline{i}, \quad \boldsymbol{a} \in E^3, \quad \text{but} \quad \underline{i}a = -a\underline{i}, \quad a \in M$$

3. We emphasize again that two quite different geometrical objects $\underline{i}$ and $e_2 e_1$, which play an important role in quantum mechanics, are both such that their square in STA is equal to -1. They are both represented by the undefined imaginary number $\sqrt{-1}$ in the spinors formalism:

(a) The pseudo-scalar of M, $\underline{i}$, corresponds to the i appearing in the V–A vectors of the electroweak theory (and, we repeat, in the expression $\boldsymbol{E} + \mathrm{i}\boldsymbol{H}$ of the electromagnetic field).

(b) The bivector of M, $e_2 \wedge e_1 = e_2 e_1 = \underline{i}e_3$ corresponds to the i of the Dirac theory. The hidden presence of a bivector in this theory corresponds to the fact that the spin, or proper angular momentum, of the electron is a bivector. It has been used in this theory in the form $\gamma_2 \gamma_1$ by Sommerfeld, probably from the beginning of the year 1930 (see [54]) and also in [39]. In fact (see Appendix B.2) the bivector spin is in the form $(\hbar c/2)n_2 \wedge n_1$, where n_1, n_2 are deduced from e_1, e_2 by the Lorentz rotation, which changes e_0 into the unit time-like vector v colinear to the Dirac current of the electron.

One of the major advantage, emphasized in [35], of the STA formalism is to avoid the confusion between these two entities, a bivector and a pseudo-scalar of M.

The identification of a Dirac spinor with a biquaternion $Q \in Cl^+(M)$ is recalled in Appendix A.2 and A.3.

We denote ψ, that we call a *Hestenes spinor*, as a biquaternion when it is applied to a wave function expressed by a Dirac spinor Ψ in Quantum Mechanics.

One can deduce the equivalences, not at all evident (Appendix A.3), established for the first time in [35]

$$\Psi \Leftrightarrow \psi, \quad \gamma^\mu \Psi \Leftrightarrow e^\mu \psi e_0, \qquad (3.16a)$$
$$i'\Psi = \Psi i' \Leftrightarrow \psi e_2 e_1, \quad i' = \sqrt{-1} \Leftrightarrow e_2 e_1 = \underline{i}e_3, \qquad (3.16b)$$

which are the key of the conversion in STA of the Dirac spinor not only in the theory of the electron but also in all the present theory of elementary particles.

The Dirac current $j \in M$ associated with a Dirac spinor Ψ is given (Appendix A.3) by the equivalence

$$j^{\mu} = \bar{\Psi}\gamma^{\mu}\Psi \in R \;\Leftrightarrow\; j = j^{\mu}e_{\mu} = \psi e_0 \bar{\psi} \in M. \tag{3.17}$$

3.2 The Hestenes Real Form of the Dirac Equation

For avoiding all ambiguity concerning the charge of the electron (see [43], p. 98) in the presentation of the Dirac equation, we denote by $q = -e$ $(e > 0)$ the charge of the electron.

Using (3.16a,b) in which ψ means the wave function of the electron, one can pass immediately from the Dirac equation ([47], (43.1))

$$\hbar c \gamma^{\mu} \partial_{\mu}(i'\Psi) - mc^2\Psi - qA_{\mu}\gamma^{\mu}\Psi = 0, \;\; i' = \sqrt{-1}, \;\; q = -e, \; (e > 0), \tag{3.18}$$

where $\partial_{\mu} = \frac{\partial}{\partial x^{\mu}}$, to the form given to this equation in [36], (2.15),

$$\hbar c e^{\mu} \partial_{\mu}\psi e_2 e_1 e_0 - mc^2\psi - qA\psi e_0 = 0, \;\;\; A = A_{\mu}e^{\mu} \in M. \tag{3.19}$$

One can find $e_1 e_2$ in place of $e_2 e_1$ in this equation ([35], (51.1)). These two possibilities correspond to the states "up" and "down" of the electron and are related with the orientation of the bivector spin (see Appendix B.2).

We work only with (3.19), the second equation giving comparable results.

3.3 The Dirac Equation in Real Biquaternion

The use of $Cl(E^3)$ instead of $Cl^+(M)$ is interesting for solving problems, in particular, the one of the central potential.

Using $e_0 e^0 = 1$, $e_0 e^k = -e_0 e_k = e_k e_0 = e_k$, $\nabla = e_k \partial_k$, we can write

$$e_0 e^{\mu} \partial_{\mu} = \partial_0 + \nabla; \;\;\; A = A^{\mu}e_{\mu} \in M, \; \boldsymbol{A} = A^k e_k \Rightarrow e_0 A = A^0 - \boldsymbol{A}. \tag{3.20}$$

Writing $e_2 e_1 = \underline{i}e_3$ in (3.19), multiplying this equation on the right and left by e_0, then on the right by $\underline{i}e_3$, we obtain

$$\partial_0\psi + \nabla\psi = -\frac{1}{\hbar c}[mc^2\bar{\psi} + q(A_0 - \boldsymbol{A})\psi]\underline{i}e_3, \;\;\; \bar{\psi} = e_0\psi e_0. \tag{3.21}$$

3.4 Notations

In the notations of Hestenes, of his followers, and in some of our own previous articles, the γ_{μ} means vectors of an orthonormal frame of M, $\sigma_k = \gamma_k\gamma_0$ means both vectors or E^3 and bivectors of M. (Note that the γ_{μ} have been already implicitly used with the above interpretation by Sommerfeld and in [39]).

For avoiding all confusion in the relations between the complex and the real formalisms, we have reserved the notations γ_{μ} and σ_k to the matrices of the complex formalism. Here $\underline{i}$ and $\underline{i}e_3$ correspond to the i and $i\sigma_3$ of Hestenes, respectively.

4

The Solutions of the Dirac Equation
for the Central Potential in the Real Formalism

Abstract. This chapter concerns a presentation of the Darwin solutions of the Dirac equation, in the Hestenes form of this equation, for the central potential problem. The passage from this presentation to that of complex spinor is entirely explicited. The nonrelativistic Pauli and Schrödinger theories are deduced as approximations of the Dirac theory.

4.1 General Approach

In the STA solutions corresponding to the one of Darwin's [21], the use of field H of the Hamilton quaternions brings notable simplifications with respect to the standard presentation and a geometrical clarity, which cannot be reached in the complex formalism.

A study of the solutions in the Hestenes real formalism has been achieved in [29]. The following presentation is based on [9, 10, 48].

Simplifications. In the form we give to ψ, the use of half integers, which is a complication, is avoided. Only the integers $m \in Z$ that appear in the associated Legendre polynomials P_ℓ^m and P_ℓ^{m+1} are employed. Half integers $m' = m + 1/2$ appear, for example, in the formula implying the total angular momentum operator of the electron (see Appendix C) and will be introduced in the Zeeman-effect (Chap. 14).

But they do not intervene in the solution for the central potential and the transition currents between two states and for this reason we prefer to use the integers m.

The expression of the solution for a given state implies only the associated Legendre polynomials P_l^m, P_l^{m+1}. The use of $P_{l\pm1}^m$, $P_{l\pm1}^{m+1}$ is avoided. Furthermore, a much more, simpler way of calculation, at least for the principal states $S1/2$, $P1/2$, $P3/2$, $D3/2$, based on a relation of recurrence, is proposed.

The direct use of the frame of spherical coordinates brings important simplifications in the calculations for the cases where two states are implied, as will be seen in the second part.

Geometrical clarity. The e_3-axe conventionally chosen for the direction of magnetic field in the Zeeman experiment plays a particular role also in the Darwin solutions. So the most convenient coordinates system is the (r, θ, φ) spherical one (2.7), in which the gradient operator ∇ is of the form

$$\nabla = \boldsymbol{n}\partial_r + \frac{1}{r}\left(\boldsymbol{w}\partial_\theta + \frac{\boldsymbol{v}}{\sin\theta}\partial_\varphi\right) \tag{4.1}$$

and gives the relations, useful for what follows

$$\boldsymbol{r} \wedge \nabla = \boldsymbol{n}\left(\boldsymbol{w}\partial_\theta + \frac{\boldsymbol{v}}{\sin\theta}\partial_\varphi\right). \tag{4.2}$$

Note that

$$\underline{i} = \boldsymbol{e}_3\boldsymbol{e}_1\boldsymbol{e}_2 = \boldsymbol{e}_3\boldsymbol{u}\boldsymbol{v} = \boldsymbol{n}\boldsymbol{w}\boldsymbol{v} \tag{4.3}$$

The vectors $\{\boldsymbol{e}_3, \boldsymbol{u}, \boldsymbol{v}\}$ or $\{\boldsymbol{n}, \boldsymbol{w}, \boldsymbol{v}\}$ are present in the expressions of the E^3 space component of space–time vectors, as the Dirac current corresponding to a state or the probability transition current between two states. But, with the use of STA these vectors are also present in the expression of the wave function ψ and allow for a simple and clear construction of these currents.

In addition, there exits a simple relation between the states, associated with the quantum number $\kappa \in Z^*$. Each state may be defined by means of a vector $\boldsymbol{N}$ in a form such that

$$\kappa \leq -1, \quad \boldsymbol{N} = L(\theta)\boldsymbol{e}_3 + M(\theta)\boldsymbol{u} \quad \Longrightarrow \kappa \geq 1, \quad \boldsymbol{N} = L(\theta)\boldsymbol{n} - M(\theta)\boldsymbol{w}, \tag{4.4}$$

the functions $L(\theta), M(\theta)$ being the same for a same value of $|\kappa|$.

4.2 The Biquaternionic Form of the Solutions in Spherical Coordinates

4.2.1 A Biquaternionic System

In the case of the central potential we have $\boldsymbol{A} = 0$. We look for a solution of (3.21) in the form

$$\psi(x^0, \boldsymbol{r}) = \phi(r)\,\mathrm{e}^{-\mathrm{i}\boldsymbol{e}_3(E/\hbar c)x^0}, \tag{4.5}$$

where now ψ has the form of a element of $Cl(E^3)$, from which we deduce

$$\nabla\phi = \frac{1}{\hbar c}[-E_0\bar{\phi} + (E + V)\phi]\underline{i}\boldsymbol{e}_3, \quad E_0 = mc^2, \ \bar{\phi} = e_0\phi e_0, \ V(r) = -qA^0. \tag{4.6}$$

Writing

$$\phi = \phi_1 + \underline{i}\phi_2 \quad \Rightarrow \quad \bar{\phi} = \phi_1 - \underline{i}\phi_2, \tag{4.7}$$

where ϕ_1, ϕ_2 are Hamilton quaternions, we obtain

$$\nabla(\phi_1 + \underline{i}\phi_2) = \frac{1}{\hbar c}[-E_0(\phi_1 - \underline{i}\phi_2) + (E + V)(\phi_1 + \underline{i}\phi_2)]\underline{i}\boldsymbol{e}_3, \tag{4.8}$$

which gives the system

$$\nabla \phi_1 = \frac{1}{\hbar c}(-E_0 - E - V)\phi_2 e_3, \tag{4.9a}$$

$$\nabla \phi_2 = \frac{1}{\hbar c}(-E_0 + E + V)\phi_1 e_3. \tag{4.9b}$$

This system is equivalent to (12.4) of [5].

4.2.2 The Fundamental Quaternionic Equation

We can obtain a solution in the form

$$\phi_1 = g(r)S, \ \phi_2 = f(r)(-nSe_3), \ g(r), f(r) \in R, \ S = F(\theta, \varphi) \in H \tag{4.10}$$

associated with the quaternionic equation

$$(r \wedge \nabla)S = \lambda S, \quad \lambda = 1 + \kappa. \tag{4.11}$$

The writing $\lambda = 1 + \kappa$ corresponds to the introduction of a conventional quantum number κ.

Note that this equation has already been considered by Sommerfeld ([54], p. 272), the quaternions being expressed in a complex form, but the solutions presented here differ in the fact that they are directly expressed by means of the vectors (u, v) or $(n, -w)$, and (we recall) only the Legendre polynomials P_l^m, P_l^{m+1} are employed, the use of $P_{l\pm1}^m$, $P_{l\pm1}^{m+1}$ being avoided.

4.2.3 The Radial Differential System

The following system [27] is to be associted with (11)

$$\frac{dg}{dr} + \frac{1+\kappa}{r}g = \frac{1}{\hbar c}(E_0 + E + V)f, \tag{4.12a}$$

$$\frac{df}{dr} + \frac{1-\kappa}{r}f = \frac{1}{\hbar c}(E_0 - E - V)g \tag{4.12b}$$

(see (14.10) of [5]).

The functions $g(r)$ and $f(r)$ are called great and fine components, respectively.

In the case of the atoms, the central potential is in the form

$$A_0 = \frac{Ze}{r}, \ V = -qA^0 = \frac{e^2 Z}{r} \Rightarrow \frac{V}{\hbar c} = \frac{\alpha Z}{r}, \ \alpha = \frac{e^2}{\hbar c}, \tag{4.13}$$

where the charge $q = -e$ of the electron is expressed in e.s.u. and Ze is the charge of the nucleus.

We recall (Sect. 2.5) that if e is expressed in e.m.u., one writes $A^0 = Ze/(4\pi\epsilon_0 r)$ and so $\alpha = e^2/(4\pi\epsilon_0)$, which gives the same value $\alpha \simeq 1/137$ to the fine structure constant α.

Indeed (4.12a) is deduced from (4.9a) in the following way. We multiply on the left of (4.9a) by $\boldsymbol{n}$, which gives, with the form (4.10) of ϕ_1 and since $\boldsymbol{n}^2 = 1$, $e_3^2 = 1$,

$$\boldsymbol{n}\nabla(gS) = \frac{1}{\hbar c}(E_0 + E + V)fS.$$

Taking into account (4.2) and (4.11)

$$\boldsymbol{n}\nabla S = (\boldsymbol{n}\wedge\nabla)S = \frac{(\boldsymbol{r}\wedge\nabla)S}{r} = \frac{1+\kappa}{r}S,$$

we obtain

$$\boldsymbol{n}\nabla(gS) = \frac{dg}{dr}S + g\frac{1+\kappa}{r}S = \left(\frac{dg}{dr} + \frac{1+\kappa}{r}g\right)S.$$

Equation (4.12a) is then obtained after the division by S.

Equation (4.12b) is deduced from (4.9b) and the form (4.10) of ϕ_2, so that

$$\nabla(f\boldsymbol{n}Se_3) = \frac{1}{\hbar c}(E_0 - E - V)gSe_3$$

and the relations

$$\nabla\boldsymbol{n} = \frac{2}{r}, \quad \nabla(\boldsymbol{n}S) = (\nabla\boldsymbol{n})S - \boldsymbol{n}\nabla S = \frac{1-\kappa}{r}S, \tag{4.14}$$

which give

$$\nabla(f\boldsymbol{n}Se_3) = \left(\frac{df}{dr} + \frac{1-\kappa}{r}f\right)Se_3.$$

Equation (4.12b) is then obtained after the division by Se_3.

4.2.4 A General Biquaternionic Solution

One can look for a solution such that [9]

$$S = \boldsymbol{N}e_3\, e^{ie_3 m\varphi}, \quad \boldsymbol{N} = L(\theta)e_3 + M(\theta)\boldsymbol{u} \tag{4.15}$$

in such a way that

$$\phi = (g\boldsymbol{N}e_3 - f\underline{i}\boldsymbol{n}\boldsymbol{N})e^{ie_3 m\varphi}.$$

4.2.5 The Dirac Probability Current and the Conditions of Normalization

The Dirac probability current $j \in M$ takes the form in the real formalism (3.17) after the elimination of $\exp \eta$, where $\eta = \underline{i}e_3(m\varphi - (E/\hbar c)x^0)$ and since $e_0\underline{i} = -\underline{i}e_0$

$$j = \psi e_0\tilde{\psi} = (gNe_3 - f\underline{i}nN)(ge_3N + fNn\underline{i})e_0. \tag{4.16}$$

The elimination of $\exp \eta$ comes from the relations $\tilde{\underline{i}} = \underline{i}$, $\tilde{e}_3 = -e_3$, $\tilde{\eta} = -\eta$, $e_0\underline{i} = -\underline{i}e_0$, and $e_0\eta = \eta e_0$, from which one deduces $(\exp \eta)e_0(\exp \eta)\tilde{} = e_0$.
We obtain then

$$j = [(g^2 + f^2)N^2 + gf\underline{i}(Ne_3Nn - nNe_3N)]e_0$$

and, from (4.15), using in particular $e_3^2 = 1$, $ue_3u = -e_3$, $\underline{i}ue_3 = v$:

$$j = (j^0 + \boldsymbol{j})e_0,$$
$$j^0 = (g^2 + f^2)N^2, \tag{4.17a}$$
$$\boldsymbol{j} = 2gf[(M^2 - L^2)\sin\theta + 2LM\cos\theta]v. \tag{4.17b}$$

The conservation $\partial_\mu j^\mu = 0$ of the current j is easy to verify: $\partial_0 j^0 = 0$ and $\boldsymbol{n}\cdot\boldsymbol{v} = 0$, $\boldsymbol{w}\cdot\boldsymbol{v} = 0$, $\boldsymbol{v}\cdot\boldsymbol{u} = 0$ imply

$$\boldsymbol{n}\cdot\partial_r\boldsymbol{j} = 0, \quad \boldsymbol{w}\cdot\partial_\theta\boldsymbol{j} = 0, \quad \boldsymbol{v}\cdot\partial_\varphi\boldsymbol{j} = 0.$$

The condition of normalization of the current in the frame $\{e_\mu\}$ is

$$\int j^0(\boldsymbol{r})\,d\tau = 1 \tag{4.18}$$

or, since $\int_0^{2\pi} v\,d\varphi = 0$, the important relation (see Sect. 13.1)

$$\int j(\boldsymbol{r})e_0\,d\tau = \int (j^0(\boldsymbol{r}) + \boldsymbol{j}(\boldsymbol{r}))\,d\tau = 1. \tag{4.18'}$$

Equation (4.18) gives

$$\int_0^\infty \int_0^{2\pi} \int_0^\pi (g^2 + f^2)N^2\,r^2\sin\theta drd\varphi d\theta = 1 \text{, which implies (4.19a)}$$

$$\int_0^\infty (g^2 + f^2)r^2\,dr = 1, \tag{4.19b}$$

$$2\pi \int_0^\pi N^2\sin\theta\,d\theta = 1 \Rightarrow \int_0^\pi N^2\sin\theta\,d\theta = \frac{1}{2\pi}. \tag{4.19c}$$

4.3 The Solution of the Quaternionic Equation

4.3.1 The Differential System Implying the Angle Theta

We use (4.2) and (4.15) for the resolution of (4.11). After derivation, of the division (4.11) by $\exp(\underline{i}e_3 m\varphi)$ and then the muliplication on the right by e_3 give

$$nw\partial_\theta N + \frac{nv}{\sin\theta}[Mv + mN\underline{i}e_3] = (1+\kappa)N. \tag{4.20}$$

From $nw = e_3 u$, $nv^2 = n$, $nv\underline{i} = w$, $we_3^2 = w$, and $wue_3 = n$, we deduce [10]

$$-\frac{dL}{d\theta}u + \frac{dM}{d\theta}e_3 + \frac{1}{\sin\theta}[(1+m)Mn + mLw] = (1+\kappa)(Le_3 + Mu). \tag{4.21}$$

Denoting $N = N^m_{1+\kappa}$, we deduce by projection upon e_3, u, the system

$$(\kappa + 1 + m)L^m_{1+\kappa} = \frac{dM^m_{1+\kappa}}{d\theta} + (1+m)\cot\theta M^m_{1+\kappa}, \tag{4.22a}$$

$$(-\kappa + m)M^m_{1+\kappa} = \frac{dL^m_{1+\kappa}}{d\theta} - m\cot\theta L^m_{1+\kappa}. \tag{4.22b}$$

Suppose now that $\kappa, m \in Z$. In the case $\kappa = 0, m = 0$, a simple calculation shows that

$$L^0_1 = C, \quad M^0_1 = -C\cot\theta - \frac{K}{\sin\theta}, \tag{4.23}$$

which is an unacceptable solution, because of the singularity for $\theta = 0$.

   This gives the explanation to the fact that $\kappa = 0$ is a forbidden quantum number. One takes

$$\kappa \in Z, \quad \kappa \neq 0 \tag{4.24}$$

4.3.2 Properties of the Solutions of Equation $(r \wedge \nabla)S = \lambda S$

We denote now by S^m_λ a solution of (4.11) corresponding to $\lambda = 1+\kappa$, $(\kappa \neq 1)$ and m, with

$$N^m_{1+\kappa} = L^m_{1+\kappa}e_3 + M^m_{1+\kappa}u, \tag{4.25}$$

where $N^m_{1+\kappa}$ satisfies (4.19c). We deduce from (4.14), then from (4.11) with $\lambda = 1 - \kappa$

$$r \wedge \nabla(nS^m_{1+\kappa}e_3) = rn\nabla(nS^m_{1+\kappa})e_3 = (1-\kappa)nS^m_{1+\kappa}e_3 = (1-\kappa)S^m_{1-\kappa}$$

and so

$$S^m_{1-\kappa} = nS^m_{1+\kappa}e_3, \Rightarrow N^m_{1-\kappa} = nN^m_{1+\kappa}e_3 \Rightarrow N^m_{1+\kappa} = mN^m_{1+\kappa}e_3. \tag{4.26}$$

A consequence is the following simple passage from the solutions $\kappa \leq -1$ to the ones such that $\kappa \geq 1$.

Since $\boldsymbol{n}e_3^2 = \boldsymbol{n}$, $\boldsymbol{n}\boldsymbol{u}e_3 = -\boldsymbol{w}$, we obtain

$$\kappa \leq -1 : \quad \boldsymbol{N}_{1+\kappa}^m = L_{1+\kappa}^m e_3 + M_{1+\kappa}^m \boldsymbol{u} \Rightarrow \boldsymbol{N}_{1-\kappa}^m = L_{1+\kappa}^m \boldsymbol{n} - M_{1+\kappa}^m \boldsymbol{w}. \quad (4.26')$$

Otherwise, we deduce from the system (4.22) that we can write

$$\boldsymbol{N}_{1+\kappa}^{-m-1} = \boldsymbol{v} \times \boldsymbol{N}_{1+\kappa}^m. \quad (4.27)$$

4.3.3 Expression of the Solutions by Means of the Legendre Polynomials

We use the normalized associated Legendre polynomials

$$P_l^m(x) = (-1)^m \left[\frac{(l-m)!}{(l+m)!}\right]^{1/2} \left[\frac{2l+1}{2}\right]^{1/2} \frac{1}{2^l l!} [1-x^2]^{m/2} \frac{d^{m+l}[(x^2-1)^l]}{dx^{m+l}}.$$

They verify the two relations (see [26], p. 45)

$$\pm \frac{d}{d\theta} P_l^m(\cos\theta) + m \cot\theta P_l^m(\cos\theta)$$

$$+[(l \pm m)(l \mp m + 1)]^{1/2} P_l^{m\mp 1}(\cos\theta) = 0 \quad (4.28)$$

and the recursion one

$$[(l+m+1)(l-m)]^{1/2} P_l^{m+1}(\cos\theta) + 2m \cot\theta P_l^m(\cos\theta)$$

$$+[(l-m+1)(l+m)]^{1/2} P_l^{m-1}(\cos\theta) = 0. \quad (4.29)$$

A General Rule.

If $|m| > l$, $P_l^m = 0$. So $-l \geq m \leq l$.

(1) Case. $\kappa \leq -1$ (States $S1/2, P3/2, \ldots$), $\kappa = -(l+1)$, $l = 0, 1, \ldots$

$$L_{-l}^m = \left[\frac{l+m+1}{2\pi(2l+1)}\right]^{1/2} P_l^m(\cos\theta), \quad (4.30a)$$

$$M_{-l}^m = \left[\frac{l-m}{2\pi(2l+1)}\right]^{1/2} P_l^{m+1}(\cos\theta). \quad (4.30b)$$

(2) Case. $\kappa \geq 1$ (States $P1/2, D3/2, \ldots$): $\kappa = l, l = 1, 2, \ldots$

$$L_{l+1}^m = \left[\frac{l-m}{2\pi(2l+1)} \right]^{1/2} P_l^m(\cos\theta), \tag{4.31a}$$

$$M_{l+1}^m = - \left[\frac{l+m+1}{2\pi(2l+1)} \right]^{1/2} P_l^{m+1}(\cos\theta). \tag{4.31b}$$

Equations (4.30) and (4.31) may be immediately verified by deduction from (4.22) after replacement of κ by $-(l+1)$ and l, respectively, which leads to (4.28+) (in which m is replaced by $m+1$), and to (4.28−).

4.3.4 Expression of the Solutions by Means of a Recursion Formula

We can write ([10])

$$[(\kappa - m - 1)(\kappa + m + 1)]^{1/2} \boldsymbol{N}_{1+\kappa}^{m+1} - 2\partial_\theta \boldsymbol{N}_{1+\kappa}^m + \boldsymbol{v} \times \boldsymbol{N}_{1+\kappa}^m \tag{4.32}$$
$$- [(\kappa - m)(\kappa + m)]^{1/2} \boldsymbol{N}_{1+\kappa}^{m-1} = 0$$

associated with

$$\boldsymbol{N}_{1-\kappa}^m = n\boldsymbol{N}_{1+\kappa}^m \boldsymbol{e}_3, \quad \boldsymbol{N}_{1+\kappa}^{-m-1} = \boldsymbol{v} \times \boldsymbol{N}_{1+\kappa}^m \tag{4.33}$$

and with the relations deduced from (4.22).

- If $\kappa \leq -1$, $\kappa \leq m \leq -\kappa - 1$,

$$\Gamma_{1+\kappa} = \left[\frac{2}{4\pi} \int_0^\pi \sin^{-2\kappa-1}\theta d\theta \right]^{1/2}, \quad \boldsymbol{N}_{1+\kappa}^{-(1+\kappa)} = (-1)^{1+\kappa}\Gamma_{1+\kappa} \sin^{-1-\kappa}\theta \boldsymbol{e}_3 \tag{4.34}$$

- If $\kappa \geq 1$, $-\kappa \leq m \leq \kappa - 1$,

$$C_{1+\kappa} = \Gamma_{1-\kappa}, \quad \boldsymbol{N}_\lambda^{1-\kappa} = (-1)^{1+\kappa}C_{1+\kappa} \sin^{\kappa-1}\theta n. \tag{4.35}$$

Equation (4.32) may be immediately verified by calculating $\partial_\theta \boldsymbol{N}_{1+\kappa}^m$ by means of (4.22), then for $\kappa = -(l+1)$ and $\kappa = l$ by projections upon $\boldsymbol{e}_3$ and $\boldsymbol{u}$, which give in each case (4.29), with $m + 1$ replacing m for the projection upon $\boldsymbol{u}$.

For example, one obtains immediately from these relations (we omit a factor $[1/4\pi]^{1/2}$)

(1) States: $\kappa \leq -1$.

- States S1/2, $\kappa = -1$: $\boldsymbol{N}_0^0 = \boldsymbol{e}_3$, $\boldsymbol{N}_0^{-1} = \boldsymbol{u}$
- States P3/2, $\kappa = -2$: $\boldsymbol{N}_{-1}^1 = -[3/2]^{1/2} \sin\theta \boldsymbol{e}_3$, $\boldsymbol{N}_{-1}^{-2} = -[3/2]^{1/2} \sin\theta \boldsymbol{u}$
  $\boldsymbol{N}_{-1}^0 = [1/2]^{1/2}(2\cos\theta \boldsymbol{e}_3 - \sin\theta \boldsymbol{u})$, $\boldsymbol{N}_{-1}^{-1} = [1/2]^{1/2}(\sin\theta \boldsymbol{e}_3 + 2\cos\theta \boldsymbol{u})$.

(2) *States:* $\kappa \geq 1$. They can be immediatly deduced from the previous ones by the changes $e_3 \to n$ and $u \to -w$.

- States P1/2, $\kappa = 1$: $N_2^0 = n$, $N_2^{-1} = -w$
- States D3/2, $\kappa = 2$: $N_3^1 = -[3/2]^{1/2} \sin \theta n$, $N_3^{-2} = [3/2]^{1/2} \sin \theta w$, $N_3^0 = [1/2]^{1/2}(2 \cos \theta n + \sin \theta w)$, $N_3^{-1} = [1/2]^{1/2}(\sin \theta n - 2 \cos \theta w)$.

4.4 Solutions of the Radial Differential System for the Discrete Spectrum

4.4.1 Solutions of the System

We have used the same notations for the energy as the ones in [5] and the method of resolution followed in [43]. With $V = \alpha Z/r$ in the system (4.12), the changes

$$G = rg, \quad F = rf, \quad \rho = 2\lambda r, \quad \lambda = \frac{mc}{\hbar}\sqrt{1 - \epsilon^2}, \quad \epsilon = \frac{E}{E_0} \tag{4.36}$$

lead to the system

$$\frac{1}{[1+\epsilon]^{1/2}}\left(\frac{dG}{d\rho} + \frac{\kappa}{\rho}G\right) = \left(\frac{1}{2} + \frac{\alpha Z}{\rho}\left[\frac{1-\epsilon}{1+\epsilon}\right]^{1/2}\right)\frac{F}{[1-\epsilon]^{1/2}}, \tag{4.37a}$$

$$\frac{1}{[1-\epsilon]^{1/2}}\left(\frac{dF}{d\rho} - \frac{\kappa}{\rho}F\right) = \left(\frac{1}{2} - \frac{\alpha Z}{\rho}\left[\frac{1+\epsilon}{1-\epsilon}\right]^{1/2}\right)\frac{G}{[1+\epsilon]^{1/2}} \tag{4.37b}$$

Asymptotic Behavior. In the case where the terms in $1/\rho$ are neglected, the system gives

$$G \simeq \sqrt{1+\epsilon}\,e^{-\rho/2}, \quad F \simeq -\sqrt{1-\epsilon}\,e^{-\rho/2}. \tag{4.38}$$

We are looking for a solution in the form

$$G = \sqrt{1+\epsilon}\,\rho^\gamma\,e^{-\rho/2}(Q_1 + Q_2), \tag{4.39a}$$
$$F = \sqrt{1-\epsilon}\,\rho^\gamma\,e^{-\rho/2}(Q_1 - Q_2). \tag{4.39b}$$

Introducing the number

$$N = \frac{\alpha Z}{\sqrt{1 - \epsilon^2}}, \tag{4.40}$$

we obtain the system

$$\rho Q_1' + (\gamma + \epsilon N - \rho)Q_1 + (\kappa + N)Q_2 = 0, \tag{4.41a}$$
$$\rho Q_2' + (\kappa - N)Q_1 + (\gamma - \epsilon N)Q_2 = 0, \tag{4.41b}$$

where the derivatives are taken with respect to ρ. When $\rho = 0$, (4.41) implies

$$\kappa^2 - N^2 = \gamma^2 - \epsilon^2 N^2 \Rightarrow \gamma^2 = \kappa^2 - \alpha^2 Z^2. \tag{4.42}$$

Taking into account this relation we obtain

$$\rho Q_1'' + (2\gamma + 1 - \rho)Q_1' - (\gamma + 1 - \epsilon N)Q_1 = 0, \tag{4.43a}$$
$$\rho Q_2'' + (2\gamma + 1 - \rho)Q_2' - (\gamma - \epsilon N)Q_2 = 0. \tag{4.43b}$$

Applying (4.62) and (4.61), one can write, introducing the constants C_1, C_2,

$$Q_1 = C_1 F(\gamma + 1 - \epsilon N, 2\gamma + 1, \rho), \tag{4.44a}$$
$$Q_2 = C_2 F(\gamma - \epsilon N, 2\gamma + 1, \rho). \tag{4.44b}$$

Because $\rho = 0$ implies $Q_1 = C_1$, $Q_2 = C_2$, one deduces from (4.41)

$$C_1 = -\frac{\gamma - \epsilon N}{\kappa - N} C_2 \tag{4.45}$$

compatible with (4.42).

A function $F(A, C, \rho)$ may be reduced to a polynomial if A is taken equal to a negative integer. Let us write

$$\gamma - \epsilon N = -n', \tag{4.46}$$

where $n' \geq 0$ is an integer.

(a) If $n' = 1, 2, \ldots$, each of the two functions of (4.44) is reductible to a polynomial.

(b) If $n' = 0$, (4.42) and (4.46) give both

$$\kappa = \pm N, \quad \gamma - \epsilon N = 0.$$

If $\kappa < 0$, (4.44) shows that because $C_1 = 0$, $Q_1 = 0$ and $Q_2 = C_2$. If $\kappa > 0$, then $C_1 = -C_2$ necessarily, a relation that is to be excluded. So we obtain

$$n' = 0, 1, 2, \ldots \quad \text{for} \quad \kappa < 0, \quad n' = 1, 2, \ldots \text{for} \quad \kappa > 0.$$

The condition of normalization (4.19a) allows one to calculate (see [I, Bechert, 1930]) the constant C_2 and so C_1 and to obtain the final normalized result

$$g, f = \frac{\pm(2\lambda)^{3/2}}{\Gamma(2\gamma + 1)} \left[\frac{(1 \pm \epsilon)\Gamma(2\gamma + 1 + n')}{4N(N - \kappa)n'!} \right]^{1/2} (2\lambda r)^{\gamma-1} e^{-\lambda r}$$
$$\times [(N - \kappa)F(-n', 2\gamma + 1, 2\lambda r) \mp n'F(1 - n', 2\gamma + 1, 2\lambda r)],$$
$$E_0 = mc^2, \quad \epsilon = \frac{E}{E_0}, \quad \lambda = \frac{E_0}{\hbar c}\sqrt{1 - \epsilon^2}, \quad N = \frac{\alpha Z}{\sqrt{1 - \epsilon^2}}. \tag{4.47}$$

4.4.2 The Levels of Energy for the Discrete Spectrum

Equations (4.40), (4.42), (4.46) give

$$n' = \frac{\alpha Z \epsilon}{\sqrt{1 - \epsilon^2}} - \sqrt{\kappa^2 - \alpha^2 Z^2}, \tag{4.48}$$

from which one deduces

$$\epsilon = \frac{E}{E_0} = \left[1 + \frac{\alpha^2 Z^2}{(n' + \sqrt{(\kappa^2 - \alpha^2 Z^2)})^2} \right]^{-1/2}. \tag{4.49}$$

One considers the integer n (principal quantum number)

$$n = n' + |\kappa|. \tag{4.50}$$

Then the energy $E = \epsilon E_0 = \epsilon mc^2$ is given by

$$E(n, \kappa) = mc^2 [1 + \frac{\alpha^2 Z^2}{(n - |\kappa| + \sqrt{(\kappa^2 - \alpha^2 Z^2)})^2}]^{-1/2}. \tag{4.51}$$

The first terms of the Taylor development of the formula giving the energy $E(n, \kappa)$ of le level n, a state corresponding to the number κ, are

$$E(n, \kappa) \simeq mc^2 \left[1 - \frac{\alpha^2 Z^2}{2n^2} - \alpha^4 Z^4 \left(\frac{1}{2n^3 |\kappa|} - \frac{3}{8n^4} \right) \right]. \tag{4.52}$$

An useful formula is the following [49]. Because $mc\alpha/\hbar = 1/a$,

$$\frac{E(n, \kappa)}{\hbar} \simeq \frac{mc^2}{\hbar} - \alpha \frac{c}{a} Z^2 \left[\frac{1}{2n^2} + \alpha^2 Z^2 \left(\frac{1}{2n^3 |\kappa|} - \frac{3}{8n^4} \right) \right]. \tag{4.53}$$

Labels Given to the State According to Their Levels of Energy. E depends on Z and on the couple of numbers (κ, n). Using an half integer j so that $j + 1/2 = |\kappa|$, one denotes, in particular, the states

$$\kappa = -1, \quad j = \frac{1}{2} \; : \; nS1/2,$$

$$\kappa = -2, \quad j = \frac{3}{2} \; : \; nP3/2,$$

$$\kappa = 1, \quad j = \frac{1}{2} \; : \; nP1/2,$$

$$\kappa = 2, \quad j = \frac{3}{2} \; : \; nD1/2.$$

4.4.3 Case of the States 1S1/2, 2P1/2, and 2P3/2

For the expression of the radial functions we use the following numbers

$$\gamma = \sqrt{1 - Z^2 \alpha^2}, \; N = \sqrt{2(1 + \gamma)}, \quad \delta = \sqrt{4 - Z^2 \alpha^2}. \tag{4.54}$$

$1S1/2$:

$$g = \left[\frac{2Z}{a}\right]^{3/2} \frac{\sqrt{\gamma+1}}{\sqrt{2\Gamma(2\gamma+1)}} \exp\left(-\frac{Zr}{a}\right) \left[\frac{2Zr}{a}\right]^{\gamma-1} \tag{4.55a}$$

$$f = -\sqrt{\frac{1-\gamma}{1+\gamma}}g. \tag{4.55b}$$

$2P1/2$:

$$C = \left[\frac{2Z}{Na}\right]^{3/2} \left[\frac{2+N}{8N(N-1)(2\gamma+1)\Gamma(2\gamma+1)}\right]^{1/2},$$

$$g = C \exp\left(-\frac{Zr}{Na}\right) \left[(2\gamma+1)(N-2)\left[\frac{2Zr}{Na}\right]^{\gamma-1} -(N-1)\left[\frac{2Zr}{Na}\right]^{\gamma}\right], \tag{4.56a}$$

$$f = -\sqrt{\frac{2-N}{2+N}}C \exp\left(-\frac{Zr}{Na}\right) \left[(2\gamma+1)N\left[\frac{2Zr}{Na}\right]^{\gamma-1} - (N-1)\left[\frac{2Zr}{Na}\right]^{\gamma}\right]. \tag{4.56b}$$

$2P3/2$:

$$g = \left[\frac{Z}{a}\right]^{3/2} \frac{\sqrt{2+\delta}}{2\sqrt{\Gamma(2\delta+1)}} \exp\left(-\frac{Zr}{2a}\right) \left[\frac{Zr}{a}\right]^{\delta-1}, \tag{4.57a}$$

$$f = -\sqrt{\frac{2-\delta}{2+\delta}}g. \tag{4.57b}$$

4.4.4 Note: The Gamma and the Confluent Hypergeometric Functions

The Gamma Function

The Gamma function $\Gamma(z)$ of the complex variable z is defined by the integral

$$\Gamma(z) = \int_0^\infty e^{-\rho}\rho^{z-\rho}\mathrm{d}\rho, \quad \mathrm{Re}(z) > 0 \tag{4.58}$$

and satisfies the property

$$\Gamma(z+1) = \Gamma(z)z$$

in such a way that if p is an integer

$$\Gamma(z+p) = \Gamma(z)z(z+1)\cdots(z+p-1). \tag{4.59}$$

In particular, since $\Gamma(1) = 1$

$$\Gamma(p) = (p-1)\,! \tag{4.60}$$

The Confluent Hypergeometric Functions

The confluent hypergeometric function $F(A, C, z)$ of the complex variable z is defined by the series

$$F(A, C, z) = 1 + \frac{A}{C}\frac{z}{1!} + \frac{A(A+1)}{C(C+1)}\frac{z^2}{2!} + \cdots \tag{4.61}$$

so that C cannot be equal to 0 or a negative integer. This function is convergent for all value of z and so defines an analytic function upon all the complex plane. If A is a negative integer $-p$, the series is reduced to a polynomial of degree p.

The function $w = F(A, C, x)$ satisfies the differential equation

$$z\frac{d^2w}{dz^2} + (C - z)\frac{dw}{dz} - Aw = 0 \tag{4.62}$$

(see [45], p. 268).

4.5 Solutions in the Pauli Approximation and for the Schrödinger Equation

4.5.1 The Pauli Approximation

In the case of a central potential Ze/r, one can look for the approximation

$$2mc^2 + \left(W + \frac{Ze^2}{r}\right) \simeq 2mc^2, \quad W = E - mc^2 \tag{4.63}$$

in such a way that the system equation (4.12a) is changed into

$$\frac{dg}{dr} + \frac{1 + \kappa}{r}g = \frac{2mc}{\hbar}f, \tag{4.64a}$$

$$\frac{df}{dr} + \frac{1 - \kappa}{r}f = -\frac{1}{\hbar c}\left(W + \frac{Ze^2}{r}\right)g \tag{4.64b}$$

We have

$$\kappa(\kappa + 1) = l(l + 1) \tag{4.65}$$

for the cases $\kappa = l$ as well as $\kappa = -(l + 1)$. The elimination of f gives the equation

$$\frac{d^2g}{dr^2} + \frac{2}{r}\frac{dg}{dr} + \left[\frac{2m}{\hbar^2}\left(W + \frac{Ze^2}{r}\right) - \frac{\kappa(\kappa + 1)}{r^2}\right]g = 0. \tag{4.66}$$

Taking into account (4.65), this equation is similar to the one of the radial solution for the Schrödinger equation (see [5], (1.12), also, in a clearer use of the physical constants, [47], IX, (19), XI, (4)).

Because $\hbar/mc = \alpha a$, (4.64a) shows that f is in order of αZ with respect to g. If one considers that f^2 is negligible with respect to g^2 because in order $(\alpha Z)^2$, the condition of normalization (4.19a) is reduced to

$$\int_0^\infty gr^2 \, \mathrm{d}r = 1 \qquad (4.67)$$

in such a way that g is exactly the same as for the Schrödinger equation.

We deduce from (4.52) and (4.53) that the value of the energy E is to be considered as reduced in such a way that

$$E \simeq mc^2 \left[1 - \frac{\alpha^2 Z^2}{2n^2}\right], \quad \text{and so} \quad \frac{2mW}{\hbar^2} = -\frac{1}{n^2}\left[\frac{Z}{a}\right]^2. \qquad (4.68)$$

Equation (4.65) becomes

$$\frac{\mathrm{d}^2 g}{\mathrm{d}r^2} + \frac{2}{r}\frac{\mathrm{d}g}{\mathrm{d}r} + \left[-\frac{1}{n^2}\left[\frac{Z}{a}\right]^2 + \frac{2}{r}\left[\frac{Z}{a}\right] - \frac{\kappa(\kappa+1)}{r^2}\right] g = 0. \qquad (4.69)$$

The Pauli–Schrödinger Theory

Using system (4.64) associated with (4.66) and (4.67) may be called the Pauli–Schrödinger theory of the electron. In this approach, the solutions are the same as for the Dirac theory, except that the radial system is now defined by (4.64), (4.66), and (4.67), with an energy E given by (4.68).

The use of this approximation of the Dirac theory is interesting for the study of the transitions in the discrete spectrum, also in the photoeffect for states of the continuum whose energy is close to mc^2, but it is no longer acceptable for the states of high energy.

When the Pauli–Schrödinger theory will be used, the states $S1/2, P1/2, \ldots$ will be denoted as $s1/2, p1/2, \ldots$.

4.5.2 Solution of the Schrödinger Equation

To be in agreement with the Schrödinger equation, where the number $\sqrt{-1}$ is to be replaced by $\underline{i}e_3$, the term implying $\boldsymbol{u}$ must disappear, the vector $\boldsymbol{N}$ must be in the form

$$\boldsymbol{N} = L(\theta)e_3$$

and the wave function becomes

$$\psi(x^0, \boldsymbol{r}) = \phi(\boldsymbol{r}) \, \mathrm{e}^{-\underline{i}e_3(E/\hbar c)x^0}, \quad \phi = g\boldsymbol{N}e_3 \, \mathrm{e}^{\underline{i}e_3 m\varphi}. \qquad (4.70)$$

The conditions of normalization are then

$$\int_0^\infty g^2 r^2 \, \mathrm{d}r = 1, \quad 2\pi \int_0^\pi L^2 \sin\theta \mathrm{d}\theta = 1 \Rightarrow L(\theta) = \frac{P_l^m(\cos\theta)}{\sqrt{2\pi}}. \qquad (4.71)$$

A detailed study of the consistency in the formulation of the Dirac, Pauli, and Schrödinger theories is achieved in [30].

4.5.3 Case of the States s1/2, p1/2, and p3/2

The function g being deduced from (4.68), the function f is then

$$s1/2: \ \kappa = -1, \ f = \frac{\alpha a}{2} \frac{dg}{dr}, \tag{4.72}$$

$$p1/2: \ \kappa = 1, \ f = \frac{\alpha a}{2} \left(\frac{dg}{dr} + \frac{2}{r} g \right), \tag{4.73}$$

$$p3/2: \ \kappa = -2, \ f = \frac{\alpha a}{2} \left(\frac{dg}{dr} - \frac{1}{r} g \right). \tag{4.74}$$

Case of the States 1s1/2, 2p1/2, and 2p3/2

The functions g, f may be obtained by two different methods giving the same results.

1. The function g is calculated as a solution of (4.69) and f is deduced from the corresponding equation above.
2. The functions g, f are obtained by neglecting in (4.55)–(4.57) $Z^2\alpha^2$ with respect to unity. In particular, we can write

$$1 + \gamma \simeq 2, \ \sqrt{1 - \gamma} \simeq \frac{Z\alpha}{\sqrt{2}}, \ N \simeq 2, \ \sqrt{2 - N} \simeq \sqrt{2 - \delta} \simeq \frac{Z\alpha}{2}$$

and use furthermore the relation $\Gamma(n) = (n-1)!$ when n is an integer.

We obtain 1s1/2:

$$g = \left[\frac{Z}{a} \right]^{3/2} 2 \exp\left(-\frac{Zr}{a} \right), \tag{4.75a}$$

$$f = -\alpha Z \left[\frac{Z}{a} \right]^{3/2} \exp\left(-\frac{Zr}{a} \right). \tag{4.75b}$$

2p1/2:

$$g = -\left[\frac{Z}{a} \right]^{3/2} \frac{1}{2\sqrt{6}} \exp\left(-\frac{Zr}{2a} \right) \frac{Zr}{a}, \tag{4.76a}$$

$$f = \alpha Z \left[\frac{Z}{a} \right]^{3/2} \frac{1}{8\sqrt{6}} \exp\left(-\frac{Zr}{2a} \right) \left(6 - \frac{Zr}{a} \right). \tag{4.76b}$$

2p3/2:

$$g = \left[\frac{Z}{a} \right]^{3/2} \frac{1}{2\sqrt{6}} \exp\left(-\frac{Zr}{2a} \right) \frac{Zr}{a}, \tag{4.77a}$$

$$f = -\alpha Z \left[\frac{Z}{a} \right]^{3/2} \frac{1}{8\sqrt{6}} \exp\left(-\frac{Zr}{2a} \right) \frac{Zr}{a}. \tag{4.77b}$$

Part II

Fields Created by the Dirac
Transition Currents
Between Two States

5

The Dirac Transition Currents Between Two States

Abstract. This chapter is devoted to the form of transition currents between two states. One can remark that, independent of the choice, real or complex, of the initial formalism, all that follows is placed in the real geometry of space–time.

5.1 Assumptions on the Source Current and the Release of Energy

5.1.1 Assumptions on the Source Current

As an imperative necessity, the source current must express that the charge of the electron associated with the state 2 is entirely found again as associated with the state 1 after the transition.

So the probability current is to be conservative for the three *consecutive* situations: state 2, transition $2 \to 1$, and state 1. Furthermore, it must correspond to a solution of the Dirac equation, which satisfies the *exterior* problem for the periods that concern the states 2 and 1.

The current $j \in M$ is deduced from the wave functions ψ_1, ψ_2, which are the solutions, *normalized to unity*, of the *exterior problem*, corresponding to the states 1 and 2.

The current j is decomposed into three part, j_2, j_{12}, j_1 corresponding to the three (successive) periods: state 2, transition $2 \to 1$, state 1

$$j = j_2 + j_{12} + j_1, \qquad (5.1)$$

where

$$j_k = \psi_k e_0 \tilde{\psi}_k, \quad \int j_k^0(\boldsymbol{r}) d\tau = 1, \quad k = 1, 2, \qquad (5.2)$$

$$j_{12} = \psi_1 e_0 \tilde{\psi}_2 + \psi_2 e_0 \tilde{\psi}_1. \qquad (5.3)$$

We emphasize that because the two situations, state 2, state 1, are *successive*, the right hand side of (5.3) is not to be multiplied by $1/2$, as in the case where the two situations would to be considered as simultaneous.

We are able to know what happens to an electron bound in an atom only by means of the long-range part of the electromagnetic field created by the electron.

This part is null for the two currents j_1 and j_2, because these currents do not depend on x^0 and that explains why an electron does not radiate *at large distance* during a stationary state. The period of time of the transition $2 \rightarrow 1$ is the only one that is able to manifest its existency.

Note that (see Appendix E)

$$\int j_{12}^0(\boldsymbol{r}) \, d\tau = 0. \tag{5.4}$$

5.1.2 Assumptions on the Release of Energy

After the assumptions that have been made about the current, the only way of calculation of the energy E released by one electron during the transition is given by the relation

$$E = E_2 - E_1, \tag{5.5}$$

where E_2 and E_1 are the levels of energy associated with the state 2 and 1.

In our elementary presentation of the transitions, we will not take into account the small correction to each of these levels called the Lamb shift. So, we adopt here for E_1 and E_2 the values given for each state 1 and 2 by the bare solutions of the Dirac equation in the exterior problem.

5.2 The Transition Current Between Two States

We consider two states 1 and 2 of energy levels E_1 and E_2 and magnetic numbers m_1 and m_2. Using (4.5) in which E is replaced by E_k and (4.15), we denote

$$\phi_k = (g_k \boldsymbol{N}_k \boldsymbol{e}_3 - f_k \underline{i} \boldsymbol{n} \boldsymbol{N}_k)^{\mathrm{i} \boldsymbol{e}_3 m_k \varphi}, \quad \boldsymbol{N}_k = L_k \boldsymbol{e}_3 + M_k \boldsymbol{u} \tag{5.6}$$

and also

$$E = E_2 - E_1, \quad \epsilon = m_1 - m_2 = -1, 0, 1, \quad \omega = (E_2 - E_1)/\hbar c. \tag{5.7}$$

We deduce from (4.5), (4.15) and (5.3) that the transition current between two states is of the form (see Appendix E)

$$j_{12} = \cos(\epsilon\varphi + \omega x^0)j_I + \sin(\epsilon\varphi + \omega x^0)j_{II}. \tag{5.8}$$

This current verifies the conservation of the charge: $e^\mu \cdot \partial_\mu j_{12} = 0$ (see Appendix F). We are interested only in the spatial component of j_{12} in the form (see Appendix E)

$$j = \cos \omega\, x^0 j_1 + \sin \omega\, x^0\, j_2, \tag{5.9}$$

$$j_1 = \cos \epsilon\varphi\, j_I + \sin \epsilon\varphi\, j_{II}, \quad j_2 = -\sin \epsilon\varphi\, j_I + \cos \epsilon\varphi\, j_{II}, \tag{5.10}$$

and (see Appendix E)

$$j_I = b(r,\theta)\, v, \quad j_{II} = a(r,\theta)\, u + c(r,\theta)\, e_3, \tag{5.11}$$

where

$$a(r,\theta) = 2[(f_1 g_2 - g_1 f_2)(L_1 L_2 + M_1 M_2) \sin\theta \\ + (g_1 f_2 + f_1 g_2)(L_1 M_2 - M_1 L_2) \cos\theta],$$

$$b(r,\theta = 2(g_1 f_2 + f_1 g_2)[(L_1 M_2 + M_1 L_2) \cos\theta \\ + (M_1 M_2 - L_1 L_2) \sin\theta],$$

$$c(r,\theta) = 2[(f_1 g_2 - g_1 f_2)(L_1 L_2 + M_1 M_2) \cos\theta \\ - (g_1 f_2 + f_1 g_2)(L_1 M_2 - M_1 L_2) \sin\theta].$$

The functions $a(r,\theta)$ and $c(r,\theta)$ are deduced from

$$j_{II} = P(r,\theta)n + Q(r,\theta)w, \tag{5.12}$$

with

$$P(r,\theta) = 2(f_1 g_2 - g_1 f_2)(L_1 L_2 + M_1 M_2),$$
$$Q(r,\theta) = 2(f_1 g_2 + g_1 f_2)(L_1 M_2 - M_1 L_2).$$

6

The Field at Large Distance Created by the Transition Currents

Abstract. This chapter concerns the emitted light and its polarization in the absence of an external field.

6.1 Polarization of the Emitted Light

We are going to calculate the vector

$$U = \cos \omega x^0 \, U_1 + \sin \omega x^0 \, U_2, \quad U_k = \int j_k(r) \, d\tau. \tag{6.1}$$

The vector (2.5)

$$E(x^0, R) = e \frac{\omega}{R} [\sin \omega(x^0 - R) \, U_1^\perp + \cos \omega(x^0 - R) \, U_2^\perp], \tag{6.2}$$

where $U_k^\perp$ is orthogonal to R is the field created by the current at a large distance R from the source.

So, the behavior of the vector U allows one to define the polarization of the emitted light.

(a) Linear polarization. U is parallel to the common direction of the polar axis e_3 chosen in (4.5) in the solutions giving each state, and time-sinusoidal (see Sect. 2.3.1).

(b) Circular polarization. U describes a time-periodic circular motion in the plane (e_1, e_2) (see Sect. 2.3.2).

We use the relations (see [5], (A.22), (A.20))

$$\cos \theta P_l^m = \left[\frac{(l+1-m)(l+1+m)}{(2l+1)(2l+3)} \right]^{1/2} P_{l+1}^m$$
$$+ \left[\frac{(l+m)(l-m)}{(2l+1)(2l-1)} \right]^{1/2} P_{l-1}^m, \tag{6.3}$$

$$\sin\theta P_l^m = -\left[\frac{(l+1+m)(l+2+m)}{(2l+1)(2l+3)}\right]^{1/2} P_{l+1}^{m+1}$$
$$+\left[\frac{(l-m)(l-1-m)}{(2l+1)(2l-1)}\right]^{1/2} P_{l-1}^{m+1}, \tag{6.4}$$

6.2 The Forbidden and Allowed Transitions

We can easily deduce from (6.3) and (6.4), and the properties of orthonormality of the associated Legendre polynomials,

$$\int_0^\pi P_j^\mu P_k^\mu \sin\theta d\theta = \delta_{jk}$$

that the vectors j_I and j_{II} are nonnull, and so the transition is observable at large distance, only in the case where the l parameters of the two states, used in the application of (4.30) and (4.31), differ only from unity. Then the transition is called "allowed." Otherwise, the transition is called "forbidden." However, additional forbidden transitions appear in what follows.

As an example, the transitions of the states $P1/2(\kappa = 1)$ and $P3/2(\kappa = -2)$ to the states $S1/2(\kappa = -1)$ are allowed transitions.

However, among the forbidden transitions, some transitions may be considered (as transitions of states $\kappa = 2$ and $\kappa = -3$ to the state $S1/2$), with a more complicated definition as the one of the allowed transitions. But their incidence is weak and their study is outside our elementary presentation.

6.3 Linear Polarization

We suppose $m_1 = m_2 = m$, and if $l_1 = l$ then $l_2 = l + 1$.

Only the component of j_{II} on e_3 intervenes in the calculation of U. We use (2.11) with the value of $c(r, \theta)$ given by (5.11) earlier. We have then to calculate with help of the relations (6.3), (6.4)

$$c_1 = 2\pi \int_0^\pi 2(L_1 L_2 + M_1 M_2) \cos\theta \sin\theta\, d\theta,$$

$$c_2 = 2\pi \int_0^\pi 2(L_1 M_2 - L_2 M_1) \sin^2\theta\, d\theta,$$

and we will have

$$U_1 = 0, \quad U_2 = Ce_3,$$
$$C = (c_1 + c_2) \int_0^\infty g_1 f_2 r^2 dr + (c_2 - c_1) \int_0^\infty g_2 f_1 r^2 dr,$$
$$U = C \sin\omega x^0 e_3. \tag{6.5}$$

Several cases are to be considered. We obtain without difficulty:

1. $\kappa_1 = -(l+1)$, $\kappa_2 = l+1$,

$$c_1 + c_2 = \frac{2(1+2m)}{2l+1}, \quad c_2 - c_1 = \frac{2(1+2m)}{2l+3}.$$

2. $\kappa_1 = -(l+1)$, $\kappa_2 = -(l+2)$,

$$c_1 + c_2 = 0, \quad c_2 - c_1 = \frac{4[(l+1-m)(l+2+m)]^{1/2}}{2l+3}.$$

3. $\kappa_1 = l$, $\kappa_2 = l+1$,

$$c_1 + c_2 = -\frac{4[(l-m)(l+1+m)]^{1/2}}{2l+1} \quad c_2 - c_1 = 0.$$

4. $\kappa_1 = l$, $\kappa_2 = -(l+2)$,

$$c_1 + c_2 = 0, \quad c_2 - c_1 = 0$$

(forbidden transition). As an example, let us calculate $c_1 + c_2$ and $c_2 - c_1$ in the case 1.

For calculating $2(L_1 L_2 + M_1 M_2) \cos\theta$ we apply (4.30a) and (4.30b) on one side and, on the other, (4.31a) and (4.31b), but by replacing in these two last equations l by $l+1$. Then we use (6.3). After integration, using $\int_0^\pi P_i^\mu P_j^\mu \sin\theta \, d\theta = \delta_{ij}$ and the elimination of 2π, we obtain

$$c_1 = \frac{2(l+m+1)(l+1-m) - (l-m)(l+m+2)}{(2l+1)(2l+3)} = \frac{2(1+2m)}{(2l+1)(2l+3)}.$$

In the same way, calculating $2(L_1 M_2 - M_1 L_2) \sin\theta$ and using (6.4), we obtain

$$c_2 = \frac{2[(l+m+1)(l+m+2) - (l-m)(l+1-m)]}{(2l+1)(2l+3)]} = \frac{4(l+1)(1+2m)}{(2l+1)(2l+3)}.$$

So, after elimination of $2l+3$ and $2l+1$, respectively, we obtain the above value of $c_1 + c_2$ and $c_2 - c_1$.

6.4 Circular Polarizations

We assume that $m_1 - m_2 = \pm 1$, $m_1 = m$, $l_1 = l$, $l_2 = l+1$. Then

$$\boldsymbol{j}_1 = \cos\varphi \, \boldsymbol{j}_I \pm \sin\varphi \, \boldsymbol{j}_{II}, \quad \boldsymbol{j}_2 = \mp\sin\varphi \, \boldsymbol{j}_I + \cos\varphi \, \boldsymbol{j}_{II}.$$

$\boldsymbol{j}_I$ and only the component of $\boldsymbol{j}_{II}$ upon $\boldsymbol{u}$ intervene and we deduce that $\boldsymbol{U}_1$, $\boldsymbol{U}_2$ are in the form

$$\boldsymbol{U}_1^\pm = (J_I \pm J_{II}) \, \boldsymbol{e}_2, \quad \boldsymbol{U}_2^\mp = (\pm J_I + J_{II}) \, \boldsymbol{e}_1.$$

J_I and J_{II} are given by (2.12), with the values of $a(r,\theta)$ and $b(r,\theta)$ given by (5.11).

(1) Case $m_1 - m_2 = +1$, $m = m_1$. Using (6.3) and (6.4) we obtain

$$A = J_I + J_{II} = a_1 \int_0^\infty g_1 f_2 r^2 \, dr + a_2 \int_0^\infty g_2 f_1 r^2 \, dr, \qquad (6.6)$$

$$\boldsymbol{U}^+ = A \left(\sin \omega x^0 \, \boldsymbol{e}_I + \cos \omega x^0 \, \boldsymbol{e}_2 \right). \qquad (6.7)$$

1. $\kappa_1 = -(l + 1)$, $\kappa_2 = l + 1$,

$$a_1 = \frac{2[(l + 1 + m)(l + 1 - m)]^{1/2}}{2l + 1}, \quad a_2 = \frac{2[(l + 1 + m)(l + 1 - m)]^{1/2}}{2l + 3}.$$

2. $\kappa_1 = -(l + 1)$, $\kappa_2 = -(l + 2)$,

$$a_1 = 0, \quad a_2 = \frac{2[(l + 1 - m)(l + 2 - m)]^{1/2}}{2l + 3}.$$

3. $\kappa_1 = l$, $\kappa_2 = l + 1$,

$$a_1 = -\frac{2[(l - m)(l + 1 - m)]1/2}{2l + 1}, \quad a_2 = 0.$$

4. $\kappa_1 = l$, $\kappa_2 = -(l + 2)$,

$$a_1 = a_2 = 0$$

(forbidden transition).

(2) Case $m_1 - m_2 = -1$, $m = m_1$. Using (6.3) and (6.4) we obtain

$$B = J_{II} - J_I = b_1 \int_0^\infty g_1 f_2 r^2 \, dr + b_2 \int_0^\infty g_2 f_1 r^2 \, dr \qquad (6.8)$$

$$\boldsymbol{U}^- = B \left(\sin \omega x^0 \, \boldsymbol{e}_I - \cos \omega x^0 \, \boldsymbol{e}_2 \right) \qquad (6.9)$$

1. $\kappa_1 = -(l + 1)$, $\kappa_2 = l + 1$,

$$b_1 = -\frac{2[(l - m)(l + 2 + m)]^{1/2}}{2l + 1}, \quad b_2 = -\frac{2[(l - m)(l + 2 + m)]^{1/2}}{2l + 3}.$$

2. $\kappa_1 = -(l + 1)$, $\kappa_2 = -(l + 2)$,

$$b_1 = 0, \quad b_2 = \frac{2[(l + 2 + m)(l + 3 + m)]^{1/2}}{2l + 3}.$$

3. $\kappa_1 = l$, $\kappa_2 = l + 1$,

$$b_1 = -\frac{2[(l + 1 + m)(l + 2 + m)]1/2}{2l + 1}, \quad b_2 = 0.$$

4. $\kappa_1 = l$, $\kappa_2 = -(l + 2)$,

$$b_1 = b_2 = 0$$

(forbidden transition).

6.5 Sum Rules for the Intensities of the Emitted Light

For a fixed couple (E_1, E_2) of levels, the sums S, S^+, S^- of the squares C^2, A^2, B^2 of the modulus of the vectors U, U^+, U^- verifies $S/2 = S^+ = S^-$. These sums are calculated by taking into account all the transitions for which $m_1 - m_2 = 0, 1, -1$, respectively.

This property may be checked, for example, in the case 1, where $\kappa_1 = -(l+1)$, $\kappa_2 = l+1$, by summing upon m:

$$m_1 - m_2 = 0 : \quad \sum_{-(l+1)}^{l} (1 + 2m)^2 = 2(l+1)(2l+1)(2l+3)/3 = s,$$

$$m_1 - m_2 = +1 : \quad \sum_{-l}^{l} (l+1+m)(l+1-m) = s/2,$$

$$m_1 - m_2 = -1 : \quad \sum_{-(l+1)}^{l-1} (l-m)(l+2+m) = s/2.$$

As a consequence, in application of (2.15), the total flux of the Poynting vector is the same for each of the set of the transitions $\Delta m = 0, 1, -1$ since for the linear polarization $U_1^2 = 0$ and for the two circular polarizations $|U_1|^2 = |U_2|^2$. So C^2 on the one side and $2A^2$ and $2B^2$ on the other side interverne in (2.15) and the number of transitions per unit of time is to be considered as the same for each set.

Using the Pauli approximation of the functions g, f, one can find again the well-known coefficients (see [3], Chap. 64) but now deduced directly from an exact relativistic calculation.

Relation with the Zeeman Effect

The above properties of the transition currents are confirmed by the observation of the normal (nonrelativistic) and abnormal (relativistic) Zeemann effect (see Part V), where the levels of energy are separated for all the values of m_1 and m_2 (see (14.20)) in each state. Such an effect makes possible the observation of the electromagnetic fields due to the separation of the energies corresponding to the different values of the number m of a state associated with a given value of the number κ. For this reason the number m is called the magnetic number, though it is to be considered even in the absence of an external magnetic field.

For the same reason the different solutions of the Dirac equation corresponding to a same state have been called the Zeeman components of this state.

7

Case of the Transitions P1/2-S1/2 and P3/2-S1/2

Abstract. This chapter concerns the transitions P1/2–S1/2, P3/2–S1/2 and spontaneous emission for these states from level 2 to level 1.

7.1 General Formulas

For simplicity we consider only transitions such that the magnetic numbers m_1, m_2 of the two states verify $m_1 = m_2 = 0$. Note that the cases $m_1 = m_2 = -1$ give the same results (with a change of sign for the transitions $P1/2 - S1/2$). Then the theorem of the sum rules may be applied for the calculation of the intensities.

We denote ψ_1 the wave function of a state $S1/2$ and ψ_2 the wave function of a state $P1/2$ and $P3/2$.

(a) Transitions P1/2–S1/2. We deduce from (6.5) and Case 1 in Sect. 6.3 with $l = 0$

$$U_1 = 0, \quad U_2 = 2 \int_0^\infty \left(g_1 f_2 + \frac{1}{3} g_2 f_1 \right) r^2 \, dr \, e_3. \tag{7.1}$$

(b) Transitions P3/2–S1/2. We deduce from (6.5) and Case 2 in Sect. 6.3 with $l = 0$

$$U_1 = 0, \quad U_2 = \frac{4\sqrt{2}}{3} \int_0^\infty g_2 f_1 r^2 \, dr \, e_3. \tag{7.2}$$

7.2 The Pauli Approximation and the Schrödinger Theory

Let

$$\Psi_1 = \phi_1(\boldsymbol{r}) \, e^{-i(E_1/\hbar c)x^0}, \quad \Psi_2 = \phi_2(\boldsymbol{r}) \, e^{-i(E_2/\hbar c)x^0} \tag{7.3}$$

be the Schrödinger wave functions corresponding to two states of energy E_1 and E_2. The transition current is such that

$$j(x^0, r) = \frac{\alpha a}{i}(\Psi_2^* \nabla \Psi_1 + \Psi_1^* \nabla \Psi_2), \tag{7.4}$$

where $\alpha a = \hbar/mc$, which gives

$$j_1(r) = 0, \quad j_2(r) = \alpha a(\phi_1 \nabla \phi_2 - \phi_2 \nabla \phi_1). \tag{7.5}$$

For the case $S = \Psi_1$ and $P = \Psi_2$, we can write

$$\phi_1(r) = \frac{1}{\sqrt{4\pi}} g_1(r), \quad \phi_2(r) = \frac{\sqrt{3}\cos\theta}{\sqrt{4\pi}} g_2(r).$$

So $U_1 = 0$ and one obtains without difficulty for the U_2 vector the vector U_a such that

$$U_a = \int j_2(r)\, d\tau = \frac{\alpha a}{\sqrt{3}} \int_0^\infty (g_1 g_2' - g_2 g_1' + \frac{2}{r}g_1 g_2)r^2\, dr\, e_3. \tag{7.6}$$

Let us denote U_b and U_c as the U_2 vectors corresponding to the transitions $p1/2 - s1/2$ and $p1/2 - s1/2$, respectively. We are going to establish the important following relations [13].

$$U_b^2 = \frac{U_a^2}{3}, \quad U_c^2 = \frac{2U_a^2}{3}, \quad U_a^2 = U_b^2 + U_c^2. \tag{7.7}$$

(a) For the transitions $p1/2 - s1/2$ we can write, using (4.73) and (4.72),

$$-\alpha a \int_0^\infty g_2 g_1' r^2\, dr = -2\alpha a \int_0^\infty g_2 g_1' r^2\, dr + \alpha a \int_0^\infty g_2 g_1' r^2\, dr,$$

$$= -2\alpha a[g_2 g_1 r^2]_0^\infty + 2\alpha a \int_0^\infty g_1 \left(g_2' + \frac{2}{r}g_2\right) r^2\, dr + \alpha a \int_0^\infty g_2 g_1' r^2\, dr,$$

where $[g_2 g_1 r^2]_0^\infty = 0$, and so

$$-\alpha a \int_0^\infty g_2 g_1' r^2\, dr = 4 \int_0^\infty g_1 f_2 r^2\, dr + 2 \int_0^\infty g_2 f_1 r^2\, dr,$$

then from (7.6) and (7.1)

$$|U_a| = |\frac{1}{\sqrt{3}} \int_0^\infty (6g_1 f_2 r^2\, dr + 2g_2 f_1 r^2\, dr)| = \sqrt{3}|U_b|. \tag{7.8}$$

(b) For the transitions $p3/2 - s1/2$ we can write, using (4.74), (4.72), (7.6), (7.2)

$$\alpha a \int_0^\infty g_1 g_2' r^2\, dr = \alpha a[g_1 g_2 r^2]_0^\infty - \alpha a \int_0^\infty g_2 \left(g_1' + \frac{2}{r}g_1\right) r^2\, dr$$

from which we deduce in a same way

$$|U_a| = |\frac{\alpha a}{\sqrt{3}} \int_0^\infty 2g_2 g_1' r^2\, dr| = \frac{4}{\sqrt{3}} |\int_0^\infty g_2 f_1 r^2\, dr| = \frac{\sqrt{3}}{\sqrt{2}}|U_c|. \tag{7.9}$$

So the relations (7.7) are verified.

7.3 Spontaneous Emission

7.3.1 The Energy Balance

The fact that in the absence of all external field an electron bound in an atom may pass from an energy state to a lower one is called spontaneus emission. The passage to a higher energy state requires the presence of an incident wave.

We follow the method used in [13] based on the energy balance, quite different from that used in [1], but which gives comparable results.

Consider two states, of energies E_1 and E_2 ($E_1 < E_2$), of an electron bound in a hydrogen-like atom. Let us write

$$A = \frac{F}{E_2 - E_1}, \tag{7.10}$$

in which F is the flux, per unit of time, through a sphere of large radius, of the Poynting vector of the electromagnetic field created by the source (see Sect. 2.4), $E = E_2 - E_1$ is the energy released by the source for one transition from state 2 to state 1 and so A is the number of transitions per unit of time.

The number A of transitions per second and the mean life of a transition $T = 1/A$ may be deduced from (2.15) in the following way. Because

$$\Omega = \frac{E_2 - E_1}{\hbar} = \omega c, \quad \frac{c\omega^2 e^2}{E_2 - E_1} = \frac{E_2 - E_1}{\hbar} \cdot \frac{e^2}{\hbar c} = \alpha\Omega,$$

we obtain

$$A = \frac{\alpha\Omega}{3}(U_1^2 + U_2^2), \quad \Omega = \frac{E_2 - E_1}{\hbar}. \tag{7.11}$$

Taking into account the equality of the flux of the Pontying vector for the cases $m_1 - m_2 = -1, 0, 1$, it is sufficient to calculate the flux for $m_1 - m_2 = 0$.

7.3.2 Spontaneous Emission in the Transitions $2P1/2 - 1S1/2$ and $2P3/2 - 1S1/2$ for the Hydrogen Atom

We consider only the case of the hydrogen atom ($Z = 1$). For simplicity we do the numerical calculation with the use of the Schrödinger approximation for the transition $2p - 1s$. The calculation for the transitions $2p1/2 - 1s1/2$ and $2p3/2 - 1s1/2$ of the Pauli–Schrödinger approximation will be deduced by means of (7.7).

The common value in these approximations of Ω is given by (4.68) with the help of (4.53), with $n = 1, 2$

$$\Omega = \alpha\frac{c}{a} \cdot \frac{3}{8}. \tag{7.12}$$

(1) Transition $2p - 1s$. The number A implies U_a^2 given by (7.6) in which g_1, g_2 are given by (4.75a) and (4.76a) (or (4.77a)), respectively. We deduce from (7.11)

$$U_a^2 = \alpha^2 8 \left[\frac{2}{3}\right]^8, \quad A = \alpha^4 \frac{c}{a} \left[\frac{2}{3}\right]^8, \tag{7.13}$$

and so

$$A = 6.268 \times 10^8 \, \text{s}^{-1}, \quad T = \frac{1}{A} = 1.595 \times 10^{-9} \, \text{s} \tag{7.14}$$

in agreement with the experimental value of $T = 1.6 \times 10^{-9}$ given in [20] and the theoretical value $T = 1.596 \times 10^{-9}$ of Wiese et al. 1966, cited in this article.

(2) Transition $2p1/2 - 1s1/2$.

$$A_b = 2.089 \times 10^8 \, \text{s}^{-1}, \quad T_b = 1.479 \times 10^{-9} \, \text{s}. \tag{7.15}$$

(3) Transition $2p3/2 - 1s1/2$.

$$A_c = 4.179 \times 10^8 \, \text{s}^{-1}, \quad T_c = 0.239 \times 10^{-9} \, \text{s}. \tag{7.16}$$

Note, as a verification, that the values of A_b A_c may be found again by means of (7.8), (4.76) and (7.9), (4.77) associated with (4.75).

The relativistic value obtained in [1], we repeat by a quite different method, are $A_b = 2.088 \times 10^8$, $A_c = 4.177 \times 10^8$, giving also $T = 1.596 \times 10^{-9}$.

Part III

Interaction with Radiation

8

Interaction with an Incident Wave: The Retardation

Abstract. In this chapter, the interactions with radiation in the so-called calculation with retardation (i.e., the fact that the action of an external plane wave is taken into account) is studied. The relativistic processing of this last problem has been considered for a long time as difficult, but we think that the pure geometrical methods used here allow one to avoid a large part of the difficulties.

8.1 Matrix Element of a Transition

Part II has been devoted to the field created in a transition between two states corresponding to the levels of energies E_1, E_2, and the phenomena of spontaneous emission (in which the final level is lower), in the absence of all external action.

Now we are going to take into account the effect of a monochromatic electromagnetic wave with a propagation vector $\mathbf{k}$ of magnitude $2\pi\nu/c$ and a polarization whose direction, orthogonal to $\mathbf{k}$, will be represented by an unit vector $\mathbf{L}$.

When the light of quantum energy $h\nu$ falls on an electron, bound in an atom, whose energy is $E_1 > 0$, a quantum may be absorbed and the electron jumps into a state of energy $E_2 = E_1 + h\nu$. The energy E_1 belongs to the discrete spectrum and E_2 may belong to the discrete (bound–bound transition) or to the continuous spectrum (photoeffect).

The transition probability is related to what is called the matrix elements of the transition (see [5], (59.3)), which are related to the transition probability current between the two states ((5.7)–(5.10) of Chap. 5) and are defined by the scalars

$$D_j^{\mathbf{k},\mathbf{L}} = \frac{1}{2} \int e^{i(\mathbf{k}.\mathbf{r})} \, \mathbf{j}_j(\mathbf{r}) \, . \, \mathbf{L} \, d\tau, \quad (j = 1, 2), \tag{8.1}$$

which are *real* numbers (see (9.12) later). A justification of the role of these scalars is made in Appendix G.

Note. The factor $1/2$ is not present in the usual presentation of the matrix elements, but is present in the usual definition of the transition currents. We have introduced this factor in the above definition because it is absent in our definition (5.3) of the current. We recall that the absence of the factor $1/2$ in the definition of the transition current has been justified in Chap. 5. It appears as a necessity for the concordance of the theoretical calculation of spontaneous emission and the experimental results concerning this phenomena.

Because **L** is orthogonal to **k**, the component of $\mathbf{j}_j(\mathbf{r})$ upon the direction of **k** does not intervene in (8.1) and we can write

$$D_j^{\mathbf{k},\mathbf{L}} = \frac{1}{2} \int e^{i(\mathbf{k}.\mathbf{r})} \, \mathbf{j}_j^\perp(\mathbf{r}) \,.\, \mathbf{L} \, d\tau, \quad (j = 1, 2), \qquad (8.2)$$

where $\mathbf{X}^\perp$ is the symbol of the component of the vector **X** orthogonal to **k**.

Let us introduce the vectors

$$\mathbf{T}_j^\perp(\mathbf{k}) = \frac{1}{2} \int e^{i(\mathbf{k}.\mathbf{r})} \, \mathbf{j}_j^\perp(\mathbf{r}) \, d\tau, \quad (j = 1, 2). \qquad (8.3)$$

It is to emphasize, as we prove in Chap. 9, (9.12), that these vectors are real:

$$\frac{1}{2} \int \sin(\mathbf{k}.\mathbf{r}) \, \mathbf{j}_j^\perp(\mathbf{r}) \, d\tau = 0 \;\Rightarrow\; \mathbf{T}_j^\perp(\mathbf{k}) = \frac{1}{2} \int \cos(\mathbf{k}.\mathbf{r}) \, \mathbf{j}_j^\perp(\mathbf{r}) \, d\tau. \qquad (8.3')$$

We can write

$$D_j^{\mathbf{k},\mathbf{L}} = \mathbf{L}.\mathbf{T}_j^\perp(\mathbf{k}). \qquad (8.4)$$

Indeed, let $\mathbf{I}_1$ be an unit vector parallel to the vector $\mathbf{T}_j^\perp(\mathbf{k})$, and so orthogonal to **k**. Let $\mathbf{I}_2$ be an unit vector orthogonal both to **k** and $\mathbf{I}_1$. Because **L** is orthogonal to **k**, it intervenes in the integral only by its components $\mathbf{L}.\mathbf{I}_1$ and $\mathbf{L}.\mathbf{I}_2$. Because the integral of the component of $\exp(i\mathbf{k}.\mathbf{r}) \, \mathbf{j}_j^\perp$ upon $\mathbf{I}_2$ is null, only $\mathbf{L}.\mathbf{I}_1$ is to be taken into account, and can be put outside the integral, giving the relation (8.4).

As a consequence, we can deduce that the determination of the matrix elements is reduced to the one of the vectors $\mathbf{T}_j^\perp(\mathbf{k})$.

Nevertheless, the choice of the direction of the vector **L** is not innocent for the determination of the value of a matrix element, and in particular, **L** can be chosen in such a way that the matrix element cancels.

Otherwise, the average of $[\mathbf{L}.\mathbf{T}_j^\perp(\mathbf{k})]^2$ on all the directions of **L** may be calculated by denoting

$$\mathbf{L} = \cos\eta\,\mathbf{I}_1 + \sin\eta\,\mathbf{I}_2, \;\; \text{so that} \;\; [\mathbf{L}.\mathbf{T}_j^\perp(\mathbf{k})]^2 = [\mathbf{T}_j^\perp(\mathbf{k})]^2 \cos^2\eta,$$

and writing

$$\frac{1}{2\pi} \int_0^{2\pi} [\mathbf{L}.\mathbf{T}_j^\perp(\mathbf{k})]^2 \, d\eta = [\mathbf{T}_j^\perp(\mathbf{k})]^2 \frac{1}{2\pi} \int_0^{2\pi} \cos^2\eta \, d\eta = \frac{1}{2}[\mathbf{T}_j^\perp(\mathbf{k})]^2. \qquad (8.5)$$

Note. The Lamb shift. The vectors $\mathbf{T}_j^\perp(\mathbf{k})$ intervene also in the form

$$[\mathbf{T}_1^\perp(\mathbf{k})]^2 + [\mathbf{T}_2^\perp(\mathbf{k})]^2 \tag{8.6}$$

in the Lamb shift calculation by means of the so called *Electrodynamics energy term* W_D, which, in addition to the *Electrostatic energy term* W_S, contributes, after the correction by the mass renormalization term, to the shift (see [25,40]). The level E_1 belongs to the discrete spectrum and E_2 may belongs to the discrete or to the continuous spectrum.

So the problem of the determination of the matrix elements is exactly the same for a transition in general and for the term W_D of the Lamb shift.

8.2 The Retardation and the Dipole Approximation

The fact that the exponential is not taken into account in (8.3) is called the "electric dipole approximation" (see Sect. 9.2.4). The fact that it is taken into account is indicated for simplicity by the word "retardation" (see [5], p. 249).

For the bound–bound absorption–emission processes, the effect of the retardation is negligible, as it is well known and as that may be confirmed numerically in a precise way (see (9.38)–(9.40) later).

However, from a theoretical and also a practical point of view, several teachings may be deduced. In particular, with the retardation, the Pauli approximation of the Dirac theory is no longer in strict agreement with the Schrödinger theory, as it is the case with the dipole approximation, when for example two states $p1/2$ and $p3/2$ are considered as unified in a single state p. Such a feature has a nonnegligible incidence.

For the bound-free transitions (photoeffect), the relativistic calculation with retardation becomes an imperative necessity for the high values of the energy of the electron in the continuum.

Relativistic Expression of the Matrix Elements

Abstract. In this chapter, the interactions with radiation in the so-called calculation with retardation (i.e., the fact that the action of an external plane wave is taken into account) is studied. The relativistic processing of this last problem has been considered for a long time as difficult, but we think that the pure geometrical methods used here allow one to avoid a large part of the difficulties.

9.1 Geometrical Construction of the Vectors $\mathbf{T}_j^\perp(\mathbf{k})$

We use the method developed in [15]. It is based on the use of the spherical system or coordinates defined by (2.7) and the expression of the vectors $\mathbf{j}_1(\mathbf{r})$, $\mathbf{j}_2(\mathbf{r})$ of (5.10).

Let $(\mathbf{I}, \mathbf{J}, \mathbf{K})$ be an orthonormal positive frame such that $\mathbf{K} = \mathbf{k}/k$, $k = |\mathbf{k}|$. There is no inconvenient to place the vectors $\mathbf{K}, \mathbf{I}$ in the plane $(\mathbf{e}_3, \mathbf{e}_1)$ in such a way that

$$\mathbf{K} = \cos\theta_0\,\mathbf{e}_3 - \sin\theta_0\,\mathbf{e}_1, \quad \mathbf{I} = \sin\theta_0\,\mathbf{e}_3 + \cos\theta_0\,\mathbf{e}_1, \quad \mathbf{J} = \mathbf{e}_2. \tag{9.1}$$

Let vector $\mathbf{n}$ may be expressed in the frame $(\mathbf{I}, \mathbf{J}, \mathbf{K})$:

$$\mathbf{n} = \cos\hat{\theta}\,\mathbf{K} + \sin\hat{\theta}\,\mathbf{U}, \quad \mathbf{U} = \cos\hat{\varphi}\,\mathbf{I} + \sin\hat{\varphi}\,\mathbf{J}. \tag{9.2}$$

One obtains immediately

$$\mathbf{k}.\mathbf{r} = kr\,\cos\hat{\theta}, \cos\hat{\theta} = \mathbf{K}.\mathbf{n} = \cos\theta_0\cos\theta - \sin\theta_0\cos\theta\cos\varphi, \tag{9.3}$$

$$e_1^\perp = \cos\theta_0\mathbf{I}, \quad e_2^\perp = 0, \quad e_3^\perp = \sin\theta_0\,\mathbf{I},$$

$$\mathbf{u}^\perp = \cos\theta_0\cos\varphi\,\mathbf{I} + \sin\varphi\,\mathbf{J}, \quad \mathbf{v}^\perp = -\cos\theta_0\sin\varphi\,\mathbf{I} + \cos\varphi\,\mathbf{J}. \tag{9.4}$$

9.1.1 Integration in the Frame (e_1, e_2, e_3)

Because the integrals in the form $\int_{-\pi}^{\pi} f(\cos \varphi) \sin \varphi \, d\varphi$ are null, one deduces from (9.3), (5.10), and (9.4) that the component of $\mathbf{T}_1^{\perp}(\pm \mathbf{k})$ upon $\mathbf{I}$ and the component of $\mathbf{T}_2^{\perp}(\pm \mathbf{k})$ upon $\mathbf{J}$ are null for the three cases $\epsilon = m_1 - m_2 = 0, 1$, or -1.

We deduce immediately from (9.4), (8.3), (5.10) and (5.11)

1. For $m_1 - m_2 = 0$

$$\mathbf{T}_1^{\perp}(\pm \mathbf{k}) = \frac{1}{2}[\int e^{\pm ikr \cos \hat{\theta}} b(r, \theta) \, \cos \varphi d\tau] \, \mathbf{J}, \tag{9.5}$$

$$\mathbf{T}_2^{\perp}(\pm \mathbf{k}) = \frac{1}{2}[\int e^{\pm ikr \cos \hat{\theta}}(a(r, \theta) \, \cos \varphi \cos \theta_0 + c(r, \theta) \, \sin \theta_0) d\tau] \mathbf{I}. \tag{9.6}$$

2. For $m_1 - m_2 = \epsilon = \pm 1$

$$\mathbf{T}_1^{\perp}(\pm \mathbf{k}) = \frac{1}{2}[\int e^{\pm ikr \cos \hat{\theta}}(b(r, \theta) \, \cos^2 \varphi + \epsilon a(r, \theta) \, \sin^2 \varphi) \, d\tau] \, \mathbf{J}, \tag{9.7}$$

$$\mathbf{T}_2^{\perp}(\pm \mathbf{k}) = \frac{1}{2}[\int e^{\pm ikr \cos \hat{\theta}}((\epsilon b(r, \theta) \, \sin^2 \varphi + a(r, \theta) \, \cos^2 \varphi) \, \cos \theta_0 +$$

$$c(r, \theta) \, \cos \varphi \sin \theta_0) d\tau] \, \mathbf{I}. \tag{9.8}$$

9.1.2 Necessity of the Integration in the Frame $(\mathbf{I}, \mathbf{J}, \mathbf{K})$

Because of the presence of $\cos \varphi$, $\cos^2 \varphi$, $\sin^2 \varphi$, and the absence of $\sin \varphi$ in the integrand, the integration in the frame (e_1, e_2, e_3) is not possible and we have to make the integration in the frame $(\mathbf{I}, \mathbf{J}, \mathbf{K})$ by using $d\tau = r^2 \sin \hat{\theta} \, d\hat{\theta} \, d\hat{\varphi} \, dr$, with the help in particular of the formulas deduced from (9.1) and (9.2):

$$\cos \theta = \mathbf{n}.\mathbf{e}_3 = \cos \hat{\theta} \cos \theta_0 + \sin \hat{\theta} \sin \theta_0 \cos \hat{\varphi}, \tag{9.9}$$

$$\sin \theta \cos \varphi = \mathbf{n}.\mathbf{e}_1 = -\cos \hat{\theta} \sin \theta_0 + \sin \hat{\theta} \cos \theta_0 \cos \hat{\varphi}. \tag{9.10}$$

Introducing the spherical Bessel functions

$$j_0(\rho) = \frac{\sin \rho}{\rho}, \quad j_2(\rho) = \left(-\frac{1}{\rho} + \frac{3}{\rho^3}\right) \sin \rho - \frac{3}{\rho^2} \cos \rho, \tag{9.11}$$

we can write

$$\int_0^{\pi} e^{\pm ikr \cos \hat{\theta}} \sin \hat{\theta} \, d\hat{\theta} = 2j_0(kr) \in R, \tag{9.12a}$$

$$\int_0^{\pi} e^{\pm ikr \cos \hat{\theta}}(1 - 3 \cos^2 \hat{\theta}) \sin \hat{\theta} \, d\hat{\theta} = 4j_2(kr) \in R. \tag{9.12b}$$

Denoting
$$F(\hat{\theta}) = 1 - 3 \cos^2 \hat{\theta}$$

and since
$$\int_0^{2\pi} d\hat{\varphi} = 2\pi, \quad \int_0^{2\pi} \cos \hat{\varphi} \, d\hat{\varphi} = 0, \quad \int_0^{2\pi} \cos^2 \hat{\varphi} \, d\hat{\varphi} = \pi,$$

we can deduce from (9.1) and (9.2) the useful relations

$$\int_0^{2\pi} \cos^2 \theta \, d\hat{\varphi} = \pi \left(\frac{2}{3} - \frac{2}{3} F(\hat{\theta}) + F(\hat{\theta}) \sin^2 \theta_0 \right) \tag{9.13}$$

$$\int_0^{2\pi} \cos \theta \sin \theta \cos \varphi \, d\hat{\varphi} = \pi F(\hat{\theta}) \cos \theta_0 \sin \theta_0, \tag{9.14}$$

$$\int_0^{2\pi} \sin^2 \theta \cos^2 \varphi \, d\hat{\varphi} = \pi \left(\frac{2}{3} + \frac{1}{3} F(\hat{\theta}) - F(\hat{\theta}) \sin^2 \theta_0 \right), \tag{9.15}$$

$$\int_0^{2\pi} \sin^2 \theta \sin^2 \varphi \, d\hat{\varphi} = \pi \left(\frac{2}{3} + \frac{1}{3} F(\hat{\theta}) \right). \tag{9.16}$$

Let $g_i(r)$ (great) and $f_i(r)$ (fine) ($i = 1, 2$) be the radial components of the Darwin solutions of the states Ψ_1, Ψ_2, we notice

$$L_{s,ij}(k) = \int_0^\infty j_s(kr) g_i(r) f_j(r) r^2 \, dr. \tag{9.17}$$

9.2 Case of the Transitions S1/2–P1/2 and S1/2–P3/2

9.2.1 Expression of the Vectors $\mathbf{T}_j^\perp(\mathbf{k})$

As an example, we detail the calculation for a transition between $\psi_2 = P1/2$ and $\Psi_1 = S1/2$, such that $\epsilon = m_1 - m_2 = 0$. The calculations for the other transitions are exactly of the same model.

Using (5.11) and also (4.30) with $l = 0$, $m_1 = 0$, for the state $S1/2$ and (4.31) with $l = 1$, $m_2 = 0$, for the state $P1/2$, or, in a much more simpler way the relations we have established in Sect. 4.2.4, we obtain immediately

$a(r, \theta) = 4 f_1 g_2 \cos \theta \sin \theta / 4\pi,$
$b(r, \theta) = 0,$
$c(r, \theta) = 2(2 f_1 g_2 \cos^2 \theta - (g_1 f_2 + f_1 g_2))/4\pi,$
and from (9.5) and (9.6)

$$\mathbf{T}_1^\perp(\pm \mathbf{k}) = 0,$$

$$\mathbf{T}_2^\perp(\pm \mathbf{k}) = \frac{1}{2} [\frac{1}{4\pi} \int e^{\pm ikr \cos \hat{\theta}} [4 f_1 g_2 \cos \theta \sin \theta \cos \varphi \cos \theta_0 +$$

$$2(2 f_1 g_2 \cos^2 \theta - (g_1 f_2 + f_1 g_2)) \sin \theta_0] d\tau] \, \mathbf{I} \tag{9.18}$$

and applying (9.14) and (9.13),

$$\mathbf{T}_2^{\perp}(\pm\mathbf{k}) = \frac{1}{4}\left[\int_0^{\infty}\left[\int_0^{\pi} e^{\pm ikr\cos\hat{\theta}}\left[2f_1g_2 F(\hat{\theta})\cos^2\theta_0 \sin\theta_0\right.\right.\right.$$

$$+\left(2f_1g_2\left(\frac{2}{3}-\frac{2}{3}F(\hat{\theta})+F(\hat{\theta})\sin^2\theta_0\right)\right.$$

$$\left.\left.\left.- 2(g_1f_2 + f_1g_2)\right)\sin\theta_0\right]\sin\hat{\theta}\,d\hat{\theta}\right]r^2dr\right]\mathbf{I} \qquad (9.19)$$

and from (9.12) and (9.17), we deduce finally the relation (9.21).

1. *Transitions* $P1/2 - S1/2$: $\kappa_1 = -1, l_1 = 0$, $\kappa_2 = 1, l_2 = 1$.
   (a) $m_1 = m_2 = 0$, $(\varepsilon = 1)$ and $m_1 = m_2 = -1$, $(\varepsilon = -1)$.

$$\mathbf{T}_1^{\perp}(\pm\mathbf{k}) = 0, \qquad (9.20)$$

$$\mathbf{T}_2^{\perp}(\pm\mathbf{k}) = \frac{\varepsilon}{3}[(3L_{0,12} + L_{0,21} - 2L_{2,21})(k)] \sin\theta_0\,\mathbf{I}. \qquad (9.21)$$

   (b) $m_1 = 0$, $m_2 = -1$, $(\varepsilon = 1)$ and $m_1 = -1$, $m_2 = 0$, $(\varepsilon = -1)$.

$$\mathbf{T}_1^{\perp}(\pm\mathbf{k}) = \frac{\varepsilon}{3}[(3L_{0,12} + L_{0,21} - 2L_{2,21})(k)]\,\mathbf{J}, \qquad (9.22)$$

$$\mathbf{T}_2^{\perp}(\pm\mathbf{k}) = \frac{1}{3}[(3L_{0,12} + L_{0,21} - 2L_{2,21})(k)]\cos\theta_0\,\mathbf{I}. \qquad (9.23)$$

2. *Transitions* $P3/2 - S1/2$: $\kappa_1 = -1, l_1 = 0$, $\kappa_2 = -2, l_2 = 1$.
   (a) $m_1 = m_2 = 0$, $(\varepsilon = 1)$ and $m_1 = m_2 = -1$ $(\varepsilon = -1)$.

$$\mathbf{T}_1^{\perp}(\pm\mathbf{k}) = -\varepsilon\frac{3}{\sqrt{2}}[(L_{2,12} + L_{2,21})(k)]\cos\theta_0 \sin\theta_0\,\mathbf{J}, \qquad (9.24)$$

$$\mathbf{T}_2^{\perp}(\pm\mathbf{k}) = \frac{\sqrt{2}}{6}[(4L_{0,21} - 3L_{2,12} + L_{2,21})(k)]\sin\theta_0\,\mathbf{I}. \qquad (9.25)$$

   (b) $m_1 = 0$, $m_2 = -1$, $(\varepsilon = +1)$ and $m_1 = -1$, $m_2 = 0$, $(\varepsilon = -1)$.

$$\mathbf{T}_1^{\perp}(\pm\mathbf{k}) = \frac{\sqrt{2}}{3}\left[(L_{0,21} - 3L_{2,12} - 2L_{2,21} + \frac{9}{2}\sin^2\theta_0(L_{2,12} + L_{2,21}))(k)\right]\mathbf{J}, \qquad (9.26)$$

$$\mathbf{T}_2^{\perp}(\pm\mathbf{k}) = \varepsilon\frac{\sqrt{2}}{3}[(L_{0,21} - 3L_{2,12} - 2L_{2,21})(k)]\cos\theta_0\,\mathbf{I}. \qquad (9.27)$$

   (c) $m_1 = 0$, $m_2 = 1$, $(\varepsilon = +1)$ and $m_1 = -1$, $m_2 = -2$, $(\varepsilon = -1)$.

$$\mathbf{T}_1^{\perp}(\pm\mathbf{k}) = \frac{\sqrt{2}}{\sqrt{3}}\left[(L_{0,21} + L_{2,21} - \frac{3}{2}\sin^2\theta_0(L_{2,12} + L_{2,21}))(k)\right]\mathbf{J}, \qquad (9.28)$$

$$\mathbf{T}_2^{\perp}(\pm\mathbf{k}) = -\varepsilon\frac{\sqrt{2}}{\sqrt{3}}[(L_{0,21} + L_{2,21})(k)]\cos\theta_0\,\mathbf{I}. \qquad (9.29)$$

9.2.2 The Relativistic Retardation and the Sum Rules

We can establish a rule similar to the rule of the sum of intensities in the normal [5], p. 254) and the anomal (Sect. 6.5) Zeemann components of the transitions between two levels of energy E_1, E_2.

We consider the following number

$$S = \frac{1}{4\pi} \int_0^{2\pi} \int_0^{\pi} ([\mathbf{T}_1^\perp(\mathbf{k})]^2 + [\mathbf{T}_2^\perp(\mathbf{k})]^2) \sin\theta_0 \, \mathrm{d}\varphi_0 d\theta_0 \qquad (9.30)$$

corresponding to an average over all the directions of the vector $\mathbf{k}$ for a given value of k, and for two given states Ψ_1, Ψ_2.

It has been established in [8] that, in the Schrödinger theory, for a fixed couple (E_1, E_2) of levels, the values of S are the same in the three cases $m_1 - m_2 = 0$, $m_1 - m_2 = 1$, $m_1 - m_2 = -1$.

This property may be extended [15] to the Dirac theory in the following way. Let us denote $S(m_1, m_2)$ as the value of S corresponding to m_1, m_2.

(1) For the transitions $S1/2 - P1/2$ one has

$$S(0,0) + S(-1,-1) = S(0,-1) = S(-1,0). \qquad (9.31)$$

This can be seen on the relations (9.22), (9.23): the integrations over φ_0 and θ_0 introduce a factor respectively equal to 4π and, because of the presence of $\cos^2\theta_0$ in the integrand, $4\pi/3$, and so a factor equal to $16\pi/3$ for the calculation of S. One observes on (9.20), (9.21) that the factor is, because of the presence of $\sin^2\theta_0$ in the integrand, $8\pi/3$, that is, the half. Since $S(0,0) = S(-1,-1)$, (9.31) is verified.

(2) For the transitions $S1/2 - P3/2$ we have

$$S(0,0) + S(-1,-1) = S(0,-1) + S(-1,-2) = S(-1,0) + S(0,1). \qquad (9.32)$$

Denoting

$$U = (L_{0,21} + L_{2,21})(k), \quad V = (L_{2,12} + L_{2,21})(k), \qquad (9.33)$$

one deduces from (9.24) to (9.29)

$$S(0,0) = S(-1,-1) = \frac{8}{27}(2U^2 - 3UV) + \frac{14}{15}V^2, \qquad (9.34)$$

$$S(0,-1) = S(-1,0) = \frac{4}{27}(2U^2 - 3UV) + \frac{16}{15}V^2, \qquad (9.35)$$

$$S(0,1) = S(-1,-2) = \frac{12}{27}(2U^2 - 3UV) + \frac{12}{15}V^2. \qquad (9.36)$$

As $8 \times 2 = 4 + 12$ and $14 \times 2 = 16 + 12$, we see that the property is verified.

9.2.3 Case of the Transitions 1S1/2–2P1/2 and 1S1/2–2P3/2

We give the values for $m_1 = m_2 = 0$ of the calculation with retardation, in the cases of the transitions $1S1/2 - 2P1/2$, $1S1/2 - 2P3/2$. For simplicity we take the Pauli approximation of the radial functions:

Using (q a positive integer)

$$\int_0^\infty e^{-\lambda r} r^q \left\{ \begin{matrix} \cos \\ \sin \end{matrix} \right\} (kr) \, \mathrm{d}r = q! \left\{ \begin{matrix} \mathrm{Re} \\ \mathrm{Im} \end{matrix} \right\} \left[\left[\frac{1}{\lambda - ik} \right]^{q+1} \right] \tag{9.37}$$

and (4.75)–(4.77) we find without difficulty
$1s1/2 - 2p1/2$:

$$\mathbf{T}_2^\perp(\pm\mathbf{k}) = Z\alpha \frac{\sqrt{2}}{\sqrt{3}} \left[\frac{2}{3} \right]^4 \frac{[1 + 4\mu^2 k^2]}{[1 + \mu^2 k^2]^3} \sin \theta_0 \, \mathbf{I}, \quad \mu = \frac{2a}{3Z}. \tag{9.38}$$

$1s1/2 - 2p3/2$:

$$\mathbf{T}_1^\perp(\pm\mathbf{k}) = -Z\alpha 3\sqrt{3} \left[\frac{2}{3} \right]^4 \frac{\mu^2 k^2}{[1 + \mu^2 k^2]^3} \cos \theta_0 \sin \theta_0 \, \mathbf{J}, \quad \mu = \frac{2a}{3Z}, \tag{9.39}$$

$$\mathbf{T}_2^\perp(\pm\mathbf{k}) = Z\alpha \frac{1}{\sqrt{3}} \left[\frac{2}{3} \right]^4 \frac{[2 - \mu^2 k^2]}{[1 + \mu^2 k^2]^3} \sin \theta_0 \, \mathbf{I}, \quad \mu = \frac{2a}{3Z}. \tag{9.40}$$

If we take the value of $k = 3\alpha/8a$ suitable to the simple absorption–emission process and if we neglect the terms in α^2, one finds again the formulas of Chap. 7. So we have the confirmation that the retardation is quite negligible for these transitions.

Note that in the photoeffect (see the tables of Chap. 12), the retardation takes a considerable importance for the high values of the energy E of the continuum.

9.2.4 Comparison with the Dipole Approximation

The formulas (9.20)–(9.29) may be verified by comparison with the values they have in the dipole approximation in which k is taken equal to 0. They are related to the total transition current vectors

$$\mathbf{U}_j = \int \mathbf{j}_j(\mathbf{r}) \, \mathrm{d}\tau$$

that we have calculated in chap. 6 for all relativistic transition.

In fact, the dipole approximation corresponds to the definition of these vectors in the absence of an external field.

So $\theta_0 = \pi/2$ corresponds to the linear polarization ($m_1 - m_2 = 0$) along the $\mathbf{e}_3$ axe and $\theta_0 = 0$ to the circular polarization ($m_1 - m_2 = \pm 1$) in the plane orthogonal to $\mathbf{e}_3$.

For $\mathbf{k} = 0$ we have, since $j_0(0) = 1, j_2(0) = 0$

$$L_{0,12}(0) = \int_0^\infty g_1 f_2 r^2 \, dr, \quad L_{0,21}(0) = \int_0^\infty g_2 f_1 r^2 \, dr, \quad L_{2,12}(0) = L_{2,21}(0) = 0.$$

It is easy to verify that

$$\mathbf{U}_1 = 2\mathbf{T}_1^\perp(0), \quad \mathbf{U}_2 = 2\mathbf{T}_2^\perp(0).$$

For example, in the case where $m_1 - m_2 = 0$, replacing θ_0 by $\pi/2$ in (9.21) and (9.25), in such a way that $\mathbf{I} = \mathbf{e}_3$, and multiplying by 2, we obtain (7.1) and (7.2) for the transitions $p1/2 - s1/2$ and $p3/2 - s1/2$.

In the case $m_1 - m_2 = \pm 1$, one can make the same verification by replacing θ_0 by 0 in (9.23), (9.27) and (9.29), in such a way that $\mathbf{I} = \mathbf{e}_1$, and multiplying by 2.

Part IV

The Photoeffect

10

The Radial Functions of the Continuum

Abstract. The general formulas of the previous part are applied to the photoeffect, that is, the jumping of the electron from a bound state (limited here to 1S1/2) to a state of the continuum and in this purpose this chapter is devoted to the Darwin solution for the continuum.

10.1 Solution of the Radial System

10.1.1 General Form of the Solution

In the Darwin solution of the Dirac equation for a hydrogenic atom, the radial functions $g(r)$ (great) and $f(r)$ (fine) satisfy the system

$$\frac{d}{dr}(gr) + \frac{\kappa}{r}(gr) = \left[\frac{1+\varepsilon}{\alpha a} + \frac{\alpha Z}{r}\right](fr), \tag{10.1a}$$

$$\frac{d}{dr}(fr) - \frac{\kappa}{r}(fr) = \left[\frac{1-\varepsilon}{\alpha a} - \frac{\alpha Z}{r}\right](gr), \tag{10.1b}$$

where $\alpha = e^2/\hbar c$ is the fine structure constant and

$$\frac{1}{\alpha a} = \frac{mc^2}{\hbar c}, \quad \varepsilon = \frac{E}{mc^2}, \tag{10.2}$$

whose solution is

$$\left\{\begin{matrix} g \\ f \end{matrix}\right\} = \pm C[1 \pm \varepsilon]^{1/2} e^{-\lambda r}(2\lambda)^\gamma r^{\gamma-1} M_\pm, \tag{10.3}$$

$$M_\pm = [(-\kappa + \frac{\eta}{\varepsilon})F(\gamma - \eta, 2\gamma + 1, 2\lambda r) \pm (\gamma - \eta)F(\gamma + 1 - \eta, 2\gamma + 1, 2\lambda r)],$$

where C is a constant depending on ε (we recall that $F(a, c, z)$ means the confluent hypergeometric function of z) and where

$$\gamma = [\kappa^2 - Z^2\alpha^2]^{1/2}, \quad \lambda = \frac{[1 - \varepsilon^2]^{1/2}}{\alpha a}, \quad \eta = \frac{Z\alpha\varepsilon}{[1 - \varepsilon^2]^{1/2}} \tag{10.4}$$

(as, e.g., in [43], Vol. 4, (36.11)).

It is to be noticed that Z and the Bohr radius a are eachone always to be associated with α except in the ratio Z/a.

10.1.2 A Choice of Variable

For a level of energy E of the continuum one has $1 - \varepsilon < 0$. We denote

$$[1 - \varepsilon]^{1/2} = i[\varepsilon - 1]^{1/2}, \quad n = \frac{Z\alpha}{[\varepsilon^2 - 1]^{1/2}}. \tag{10.5}$$

We emphasize that, in no way, the definition of the real number n implies an approximation. It corresponds simply to the choice of a variable related to the energy. We deduce

$$[\varepsilon^2 - 1]^{1/2} = \frac{Z\alpha}{n}, \quad \lambda = \frac{[(1 - \varepsilon)(1 + \varepsilon)]^{1/2}}{\alpha a} = \frac{i[\varepsilon^2 - 1]^{1/2}}{\alpha a} = \frac{iZ}{na}, \tag{10.6}$$

$$\frac{\eta}{\varepsilon} = \frac{Z\alpha}{i[\varepsilon^2 - 1]^{1/2}} = -in, \quad \varepsilon = \frac{[n^2 + Z^2\alpha^2]^{1/2}}{n}. \tag{10.7}$$

Introducing the real number ν whose role is important in what follows

$$\nu = [n^2 + Z^2\alpha^2]^{1/2}, \quad \text{we have } \eta = -i\nu \tag{10.8}$$

denoting

$$\frac{\gamma - \eta}{-\kappa + \frac{\eta}{\varepsilon}} = e^{i2\xi} = \frac{\gamma + i\nu}{-\kappa - in}, \tag{10.9}$$

we obtain the radial functions in the normalization on the energy scale ε (see Sect. 10.1.3)

$$\left\{\begin{matrix} g \\ f \end{matrix}\right\} = [\varepsilon \pm 1]^{1/2} \frac{e^{\nu\pi/2}|\Gamma(\gamma + 1 - i\nu)|\sqrt{n}}{\sqrt{\pi}\sqrt{\alpha a}\sqrt{\alpha Z}\,\Gamma(2\gamma + 1)} \left[\frac{2Z}{na}\right]^{\gamma} r^{\gamma-1} \left\{\begin{matrix} \mathrm{Re} \\ \mathrm{Im} \end{matrix}\right\} S, \tag{10.10}$$

$$S = \exp\left[-i\left(\xi + \frac{Zr}{na}\right)\right] F\left(\gamma + i\nu, 2\gamma + 1, i\frac{2Zr}{na}\right) \tag{10.11}$$

This equation may be compared with the (36.19) of [43], Vol. 4), for example, by taking into account that here there is a change of sign in (10.5), an exchange between the real and the imaginary part of S due to our exchange between cosine and sinus in the asymptotic definition of g and f, that the normalization is made on the *energy scale* ε (see Sect. 10.1.3), and also that the equation cited in reference contains an useless factor 2.

We can notice that

$$[\varepsilon - 1]^{1/2} = \frac{\alpha Z}{n[\varepsilon + 1]^{1/2}}. \tag{10.12}$$

10.1.3 Normalization on the Energy Scale

Denoting

$$C_1 = C i^\gamma \left(-\kappa + \frac{\eta}{\varepsilon} \right) e^{i\xi} \qquad (10.13)$$

we obtain

$$\begin{Bmatrix} g \\ f \end{Bmatrix} = \pm C_1 \begin{Bmatrix} [\varepsilon + 1]^{1/2} \\ i[\varepsilon - 1]^{1/2} \end{Bmatrix} \left[\frac{2Z}{na} \right]^\gamma r^{\gamma - 1} N_\pm, \qquad (10.14)$$

$$N_\pm = e^{-i\frac{Z}{na}} \left[e^{-i\xi} F\left(\gamma + i\nu, 2\gamma + 1, i\frac{2Zr}{na} \right) \pm e^{i\xi} F\left(\gamma + 1 + i\nu, 2\gamma + 1, i\frac{2Zr}{na} \right) \right].$$

The asymptotic behavior.

Using the asymptotic expansion of $F(a, c, z)$ for large $|z|$ in which one applies for $R = r$ large $R^{-(\gamma+1)} \ll R^{-\gamma}$ and denoting

$$\Gamma(\gamma + 1 \mp i\nu) = |\Gamma(\gamma + 1 - i\nu)| e^{\pm i\chi}, \qquad (10.15)$$

we obtain for each $\varepsilon > 0$

$$\begin{Bmatrix} g(R, \varepsilon) \\ f(R, \varepsilon) \end{Bmatrix} \simeq \pm C_1(\varepsilon) \begin{Bmatrix} [\varepsilon + 1]^{1/2} \\ i[\varepsilon - 1]^{1/2} \end{Bmatrix} \frac{\Gamma(2\gamma + 1)}{|\Gamma(\gamma + 1 - i\nu)| R} L_\pm, \qquad (10.16)$$

$$L_\pm = \left[(-i)^{-\gamma - i\nu} \left[\frac{2ZR}{na} \right]^{-i\nu} e^{-i(\frac{ZR}{na} + \chi + \xi)} \pm i^{-\gamma - i\nu} \left[\frac{2ZR}{na} \right]^{i\nu} e^{i(\frac{ZR}{na} + \chi + \xi)} \right],$$

and because we can write

$$(-i)^{-\gamma - i\nu} = e^{i\gamma\pi/2} e^{-\nu\pi/2}, \quad i^{-\gamma + i\nu} = e^{-i\gamma\pi/2} e^{-\nu\pi/2}, \quad \frac{ZR}{na} \ll \nu \ln \frac{2ZR}{na}. \qquad (10.17)$$

We obtain

$$\begin{Bmatrix} g(R, \varepsilon) \\ f(R, \varepsilon) \end{Bmatrix} \simeq \pm C_1 \begin{Bmatrix} [\varepsilon + 1]^{1/2} \\ i[\varepsilon - 1]^{1/2} \end{Bmatrix} \frac{\Gamma(2\gamma + 1) e^{-\nu\pi/2}}{|\Gamma(\gamma + 1 - i\nu)| R} 2 \begin{Bmatrix} \cos \\ -i\sin \end{Bmatrix} \left[\frac{ZR}{na} + \phi_1 \right], \qquad (10.18)$$

with $\phi_1(\varepsilon) = \chi + \xi - \frac{\pi}{2}\gamma$.

The conservation of the charge and the asymptotic behavior on the energy scale $\varepsilon = E/mc^2$.

On the one side, the conservation of the charge is given by

$$\lim_{R \to \infty} \int_{\varepsilon_0}^{\varepsilon_1} \left[\int_0^R (g(r, \varepsilon)g(r, \varepsilon') + f(r, \varepsilon)f(r, \varepsilon')) r^2 \, dr \right] d\varepsilon' = 1, \varepsilon_0 < \varepsilon < \varepsilon_1 \qquad (10.19)$$

associated with the relation deduced of the system (10.1)

$$\int_0^R (g(r,\varepsilon)g(r,\varepsilon')+f(r,\varepsilon)f(r,\varepsilon'))r^2\,dr = \frac{f(R,\varepsilon)g(R,\varepsilon')-g(R,\varepsilon)f(R,\varepsilon')}{(\varepsilon'-\varepsilon)}\alpha a R^2,$$

$$(10.20)$$

on the otherside, by the solution of this system in which the terms in $1/r$ have been removed

$$\left\{\begin{matrix} g(R,\varepsilon) \\ f(R,\varepsilon) \end{matrix}\right\} \simeq C_2(\varepsilon)\frac{[\varepsilon\pm 1]^{1/2}}{\sqrt{\alpha a}R}\left\{\begin{matrix} \cos \\ -\sin \end{matrix}\right\}\left[[\varepsilon^2-1]^{1/2}\frac{R}{\alpha a}+\phi_2(\varepsilon)\right] \quad (10.21)$$

Replacing in (10.20) $g(R,\varepsilon)$, $f(R,\varepsilon)$ by their above values and applying (after convenient rearrangement and changes of variable) the Dirichlet theorem to the relation (10.19), we obtain

$$\pi[C_2(\varepsilon)]^2[\varepsilon^2-1]^{1/2}=1 \quad\Rightarrow\quad C_2(\varepsilon)=\frac{1}{\sqrt{\pi}[\varepsilon^2-1]^{1/4}}=\frac{\sqrt{n}}{\sqrt{\pi}\sqrt{\alpha Z}}. \quad (10.22)$$

The comparison with (10.18) allows one to identify ϕ_1 and ϕ_2 and gives

$$C_1=\frac{e^{\nu\pi/2}|\Gamma(\gamma+1-i\nu)|\sqrt{n}}{2\sqrt{\pi}\sqrt{\alpha a}\sqrt{\alpha Z}\,\Gamma(2\gamma+1)}. \quad (10.23)$$

Applying $F(a,c,z)=e^z F(c-a,c,-z)$

$$F\left(\gamma+1+i\nu,2\gamma+1,i\frac{2Zr}{na}\right)=e^{i\frac{2Z}{na}}F\left(\gamma-i\nu,2\gamma+1,-i\frac{2Zr}{na}\right), \quad (10.24)$$

we obtain

$$N_\pm=2\left\{\begin{matrix} \mathrm{Re} \\ \mathrm{Im} \end{matrix}\right\}\left[\exp\left(-i(\xi+\frac{Zr}{na})\right)F\left(\gamma+i\nu,2\gamma+1,i\frac{2Zr}{na}\right)\right] \quad (10.25)$$

and eliminating the number 2 between (10.23) and (10.25) we deduce the relation (10.10).

Note. When the normalization is made on the p-scale (see [43], Vol. 4 paragraph 36), where $p=[\varepsilon^2-1]^{1/2}$, $d\varepsilon'$ is to be replaced in (10.19) by dp', and $C_2(\varepsilon)$ is to be replaced by $\hat{C}_2(\varepsilon)$ such that

$$\pi[\hat{C}_2(\varepsilon)]^2[\varepsilon^2-1]^{1/2}\frac{dp}{d\varepsilon}=1 \quad\Rightarrow\quad \hat{C}_2(\varepsilon)=\frac{1}{\sqrt{\pi\varepsilon}}. \quad (10.26)$$

10.2 The Different Approximations of the Radial Functions

Three kinds of approximations may be considered:

10.2.1 The Approximation $Z^2\alpha^2 \ll \kappa^2$

This approximation only implies that γ is replaced by $|\kappa|$:

$$\gamma \simeq |\kappa|, \quad e^{-i\xi} \simeq e^{-i\xi_0}, \quad e^{-i\xi_0} = \left[-\frac{\kappa + in}{|\kappa| + i\nu}\right]^{1/2}. \tag{10.27}$$

That gives

$$\left\{\begin{matrix} g \\ f \end{matrix}\right\} = [\varepsilon \pm 1]^{1/2}\frac{e^{\nu\pi/2}|\Gamma(|\kappa| + 1 - i\nu)|\sqrt{n}}{\sqrt{\pi}\sqrt{\alpha a}\sqrt{\alpha Z}\,(2|\kappa|)!}\left[\frac{2Z}{na}\right]^{|\kappa|} r^{|\kappa|-1}\left\{\begin{matrix} \mathrm{Re} \\ \mathrm{Im} \end{matrix}\right\} S, \tag{10.28}$$

$$S = \exp\left[-i\left(\xi_0 + \frac{Zr}{na}\right)\right] F\left(|\kappa| + i\nu, 2|\kappa| + 1, i\frac{2Zr}{na}\right) \tag{10.29}$$

and where $\varepsilon = [n^2 + Z^2\alpha^2]^{1/2}/n$ and $\nu = [n^2 + Z^2\alpha^2]^{1/2}$ remain unchanged.

10.2.2 The Approximation $Z^2\alpha^2 \ll n^2$ or Pauli–Schrödinger Approximation

It intervernes in addition to the previous one and leads to write

$$\gamma \simeq |\kappa|, \quad \varepsilon + 1 \simeq 2, \quad [\varepsilon^2 - 1]^{1/2} \simeq [2(\varepsilon - 1)]^{1/2} = \frac{Z\alpha}{n}, \tag{10.30}$$

$$\frac{\eta}{\varepsilon} = -in \simeq \eta, \quad \nu \simeq n, \quad \xi \simeq \xi_0, \quad e^{i2\xi_0} = -\frac{|\kappa| + in}{\kappa + in}. \tag{10.31}$$

Using the relations

1. If $\kappa = \ell > 0$: $F(a, c, x) - F(a - 1, c, x) = xF(a, c + 1, x)/c$,
2. If $\kappa = -(\ell + 1) < 0$:

$$aF(a, c + 1, x) - (a + 1 - c)F(a, c, x) = (c - 1)F(a, c + 1, x),$$

it is a simple matter to show that $g(r)$ becomes the solution $R(r)$ of the Schrödinger equation on the energy scale ε (see [5]], (4.20), (4.21), (4.23)).

$$R(r) = \frac{e^{n\pi/2}\sqrt{2}|\Gamma(\ell + 1 - in)|\sqrt{n}}{\sqrt{\pi}\sqrt{\alpha a}\sqrt{\alpha Z}\,(2\ell + 1)!}\frac{Z}{na}\left[\frac{2Zr}{na}\right]^{\ell} e^{-iZr/na} F$$
$$\left(\ell + 1 + in, 2\ell + 2, i\frac{2Zr}{na}\right). \tag{10.32}$$

Thus, this second approximation corresponds to the transformation of the system (10.1) into the system of the radial functions in the Pauli approximation for which, in (10.1a) the left hand side is replaced by $(2/\alpha a)f(r)$, (10.1b) remaining unchanged.

10.2.3 The Schrödinger Approximation

This approximation lies in the use of (10.32) for the radial function associated with the form of the Schrödinger wave function.

10.2.4 Interest and Validity of the Approximations

Presently the use of computers allows the calculation of the exact relativistic formulas of the photoeffect even when the effect on the formulas of the presence of the incident wave (the "retardation") is taken into consideration. The degree of exactitude depends only on the chosen precision in the numerical calculation.

An interest of the above approximations lies in the fact that their common relation $Z^2\alpha^2 \ll \kappa^2$ allows the use of the method of Laplace for the calculation of the confluent hypergeometric functions and lead to analytic results. This method has been for a long time the only way of calculation.

Another interest is the comparison between the relativistic and the nonrelativistic approachs of the theory of the electron. These approachs give about the same results concerning the bound–bound transitions, also in the photoeffect when the energy in the continuum is close to the freedom one. The good concordance of the results obtained by the two methods for the energies of the continuum close to this energy will give a strong credit to the validity of the relativistic results for the high energies, which differ widely from the nonrelativistic ones.

The first approximation coincide with the Pauli approximation only for the discrete spectrum, but not for the continuum. One can expect that the first approximation has a weak incidence on the result, independently of the level of energy considered in the continuum, but the second one is directly related to the value of the number n in respect with $Z\alpha$ and may lead to important differences for the weak values of n, that is, the high values of the energy. So in what follows, we mainly use the first approximation, the second one being devoted only to the verification of the results, by a passage to the well known nonrelativistic expressions (see [5], Sect. 71) of the matrix elements in the dipole approximation.

11

Matrix Elements for the Transitions
1S1/2-Continuum

Abstract. This chapter concerns the transitions from the state 1S1/2 to the states
P1/2, P3/2 in the dipole approximation (i.e., the fact that the retardation is not
taken into account) and the transitions 1S-P with retardation in the Schrödinger
theory.

11.1 The transitions 1S1/2-Continuum in the Dipole and Schrödinger Approximations

All the followings approximations use the relation $Z^2\alpha^2 \ll \kappa^2$ and cannot be
applied for large values of Z.

As it is schown in Sect. 9.1, (9.17) one has to calculate integrals in the form

$$\int_0^\infty J(kr)g_1 f_2 r^2 \mathrm{d}r \quad \text{or} \quad \int_0^\infty J(kr)f_1 g_2 r^2 \mathrm{d}r, \qquad (11.1)$$

where $J(kr)$ is the spherical Bessel function $j_0(kr)$ or $j_2(kr)$ (which is reduced
to unity or zero in the case of the dipole approximation).

The radial functions g_1, f_1 corresponding to $1S1/2$ in the Pauli approximation (4.69), (4.72)

$$g_1 = \left[\frac{Z}{a}\right]^{3/2} 2e^{-Zr/a}, \quad f_1 = -\alpha Z \left[\frac{Z}{a}\right]^{3/2} e^{-Zr/a}. \qquad (11.2)$$

The functions g_2, f_2 correspond to a state of the continuum, and thus one
has to consider integrals in the form

$$I(A, C, p; k) = \int_0^\infty \exp[-(1+\frac{i}{n})\frac{Zr}{a}]J(kr)r^p F(A, C, i\frac{2Zr}{na})\mathrm{d}r, \qquad (11.3)$$

where $A = |\kappa| + i\nu$, $C = 2|\kappa| + 1$, $p = |\kappa| + 1$.

The calculation of these integral may be achieved in two different ways:

(a) The Laplace method of the representation of a confluent hypergeometric function (see Sect. 12.2).
(b) The direct integration term by term of the integrals (11.1) which may be achieved by means of (11.30) below, leading to the calculation of hypergeometric series which may be reduced here to polynomials

We will not detail here the calculations. They are explicited in [16].

11.2 Transitions 1S1/2-P1/2 in the Dipole Approximation

The interest of a calculation with the dipole approximation is to show, by comparison, the incidence of the retardation. It may be considered as negliglide for the discrete spectrum and the values of the energy E in the continuum close to the freedom energy. But this incidence becomes important and even considerable for the high values of E.

In (9.17), $j_0(kr)$ and $j_2(kr)$ are to be replaced by 1 and 0, respectively.

We will consider the case where the difference between the magnetic quantum number m_1 and m_2 is equal to zero.

One deduces from (9.20), (9.21) in which $\theta_0 = \pi/2$, and so $\mathbf{I} = \mathbf{e}_3$

$$\mathbf{T}_1^\perp(0) = \frac{1}{2}\mathbf{U}_1 = 0$$

$$\mathbf{T}_2^\perp(0) = \frac{1}{2}\mathbf{U}_2 = [\int_0^\infty g_1 f_2 r^2 dr + \frac{2}{3}\int_0^\infty f_1 g_2 r^2 dr]\,\mathbf{e}_3 \qquad (11.4)$$

in conformity with (7.1). Here $|\kappa_2| = 1$. Using the relation

$$|\Gamma(2 - i\nu)| = \sqrt{\pi}\sqrt{\nu}[1 + \nu^2]^{1/2}\left[\frac{2}{e^{\nu\pi} - e^{-\nu\pi}}\right]^{1/2} \qquad (11.5)$$

(see [5], (5.21)), replacing in (10.29) $\exp(-i\xi_0)$ by its value (10.27), eliminating $[1 + \nu^2]^{1/2}$, one obtains

$$\left\{\begin{array}{c} g_2 \\ f_2 \end{array}\right\} = [\varepsilon \pm 1]^{1/2}\left[\frac{Z}{a}\right]^{1/2}\frac{\sqrt{2}\sqrt{\nu n}}{\alpha a[1 - e^{-2\nu\pi}]^{1/2}n}\left\{\begin{array}{c} \mathrm{Re} \\ \mathrm{Im} \end{array}\right\}S \qquad (11.6)$$

$$S = i[(1 + in)(1 - i\nu)]^{1/2}e^{-iZr/na}F(1 + i\nu, 3, i\frac{2Zr}{na}) \qquad (11.7)$$

We obtain (see [16], (11.36)–(11.43))

$$|\mathbf{T}_2^\perp(0)| = \left|\frac{2\sqrt{2}\,n^2\sqrt{\nu n}\,e^{-2\nu\cot^{-1}n}}{(1 + n^2)^2[1 - e^{-2\nu\pi}]^{1/2}}\left[\frac{2}{n[\varepsilon + 1]^{1/2}}\mathrm{Im}[L] - \frac{[\varepsilon + 1]^{1/2}}{3}\mathrm{Re}[L]\right]\right|$$

$$L = -[(1 + in)(1 - i\nu)]^{1/2}(1 + in) \qquad (11.8)$$

11.3 Transitions 1S1/2-P3/2 in the Dipole Approximation

We will also consider the case $m_1 = m_2 = 0$.

We deduce from (9.24), (9.25) in which $\theta_0 = \pi/2$, and so $\mathbf{I} = \mathbf{e}_3$

$$\mathbf{T}_1^{\perp}(0) = \frac{1}{2}\mathbf{U}_1 = 0$$

$$\mathbf{T}_2^{\perp}(0) = \frac{1}{2}\mathbf{U}_2 = \frac{2\sqrt{2}}{3}\int_0^{\infty} f_1 g_2 r^2 \mathrm{d}r \ \mathbf{e}_3 \qquad (11.9)$$

in conformity with (7.2). From $|\kappa_2| = 2$ and

$$|\Gamma(3 - i\nu)| = \sqrt{\pi}\sqrt{\nu}[(1 + \nu^2)(4 + \nu^2)]^{1/2}\left[\frac{2}{e^{\nu\pi} - e^{-\nu\pi}}\right]^{1/2} \qquad (11.10)$$

replacing $\exp(-i\xi_0)$ in (10.29) by its value (10.27), eliminating $[4 + \nu^2]^{1/2}$, one obtains

$$\left\{\begin{matrix} g_2 \\ f_2 \end{matrix}\right\} = [\varepsilon \pm 1]^{1/2}\left[\frac{Z}{a}\right]^{3/2}\frac{[1 + \nu^2]^{1/2}\sqrt{\nu n}\ r}{\alpha a\ 3\sqrt{2}[1 - e^{-2\nu\pi}]^{1/2}n^2}\left\{\begin{matrix} \mathrm{Re} \\ \mathrm{Im} \end{matrix}\right\}S \qquad (11.11)$$

$$S = [(2 - in)(2 - i\nu)]^{1/2}e^{-iZr/na}F(2 + i\nu, 5, i\frac{2Zr}{na}) \qquad (11.12)$$

we obtain (see [16], (11.49)–(11.51), with a change of sign in the expression of J_1 and the correction of a priting erratum in the term $2\nu(5 + 2\nu^2)n$ lying in N below)

$$|\mathbf{T}_2^{\perp}(0)| = |2[\varepsilon + 1]^{1/2}\frac{[1 + \nu^2]^{1/2}\sqrt{\nu n}\ n^2}{[1 - e^{-2\nu\pi}]^{1/2}}\mathrm{Re}[S]|, \qquad (11.13)$$

where

$$S = [(2 - in)(2 - i\nu)]^{1/2}\frac{N}{i\nu(2 - i\nu)(1 + \nu^2)}$$

$$N = 1 - [3n^4 + 6\nu n^3 + 6(1 + \nu^2)n^2 + 2\nu(5 + 2\nu^2)n$$

$$+3(1 + 2\nu^2) - i4\nu(\nu^2 + 1)]\frac{e^{-2\nu\cot^{-1}n}}{3(1 + n^2)^2}$$

We give an indication on the method based on the formulas of the Note below, using the hypergeometric polynomials. We deduce from (11.3) that because $p = 2 + 1$, and after the change $\rho = Zr/a$, we have to calculate an integral in the form

$$L = \int_0^{\infty}\exp[-(1 + \frac{i}{n})]\rho^3 F(2 + i\nu, 5, \frac{i2}{n}\rho)\mathrm{d}\rho$$

We apply the (11.30), below which introduces the functions

$$F(2 + i\nu, 4, 5, \frac{i2}{n + i})$$

then (11.32) where appear

$$F(2 + i\nu, 4, 2 + i\nu, \frac{in + 1}{in - 1}), \ F(3 - i\nu, 1, -i\nu, \frac{in + 1}{in - 1})$$

These two functions lead, by means of (11.31), to the polynomials

$$F(0, -2 + i\nu, 2 + i\nu, \frac{in + 1}{in - 1}) = 1, \ F(-3, -1 - \nu, -i\nu, \frac{in + 1}{in - i})$$

It is easy to verify that the first above polynomial, equal to 1 (it corresponds to the residue at infinity in the integration by the Laplace method), gives, in combination with the second polynomial, and with the help of the following relation

$$\left[\frac{in - 1}{in + 1}\right]^{\nu} = e^{-2\nu \cot^{-1} n}$$

the number N.

11.4 Transitions 1s-p in the Schrödinger Theory

We follows the method of calculation of [55]. The Schrödinger waves functions ψ_1 corresponding to $1s$ and ψ_2 corresponding to p are in the form

$$\psi_1(\mathbf{r}) = \frac{2}{\sqrt{4\pi}} \left[\frac{Z}{a}\right]^{3/2} e^{-Zr/a}, \quad \psi_2(\mathbf{r}) = \frac{\sqrt{3}}{\sqrt{4\pi}} \cos\theta R(r), \tag{11.14}$$

where $R(r)$ is given by (10.32) in which $\ell = 1$, with the help of (11.5) in which $\nu = n$, so that

$$R(r) = \frac{2}{3\alpha a} \left[\frac{Z}{a}\right]^{3/2} \frac{[1 + n^2]^{1/2}}{n[1 - e^{-2n\pi}]^{1/2}} r e^{-iZr/an} F(2 + in, 4, i\frac{2Zr}{na}) \tag{11.15}$$

the definition of n with respect to the energy $E_2 = \varepsilon mc^2$ being given by (10.30).

Using the properties of the Schrödinger current we can write

$$\mathbf{T}^{\perp}(\mathbf{k}) = \frac{\hbar}{mc} \int e^{i\mathbf{k}\cdot\mathbf{r}} [\psi_2 \nabla \psi_1]^{\perp} d\tau = -\frac{\hbar}{mc} \int e^{i\mathbf{k}\cdot\mathbf{r}} [\psi_1 \nabla \psi_2]^{\perp} d\tau, \quad \frac{\hbar}{mc} = \alpha a \tag{11.16}$$

Considering the frame $\mathbf{I}, \mathbf{J}, \mathbf{K}$ where $\mathbf{K} = \mathbf{k}/k$, defined in Sect. 9.1, the relations

$$\mathbf{n}.\mathbf{e}_3 = \cos\theta = \cos\hat{\theta}\cos\theta_0 + \sin\hat{\theta}\sin\theta_0\cos\hat{\varphi}$$

$$[\nabla(e^{-Zr/a})]^{\perp} = [\mathbf{n}(e^{-Zr/a})']^{\perp} = -\frac{Z}{a}e^{-Zr/a}\sin\hat{\theta}\,\mathbf{U} \qquad (11.17)$$

$$d\tau = r^2\sin\hat{\theta}d\hat{\theta}d\hat{\varphi}dr, \qquad \int_0^{2\pi}\sin\hat{\theta}\cos\theta\,\mathbf{u}\,d\hat{\varphi} = \pi\sin^2\hat{\theta}\sin\theta_0\mathbf{I} \qquad (11.18)$$

$$\int_0^{\pi}e^{\pm ikr\cos\hat{\theta}}\sin^3\hat{\theta}d\hat{\theta} = 4J(kr), \quad J(kr) = \frac{\sin kr}{(kr)^3} - \frac{\cos kr}{(kr)^2} \qquad (11.19)$$

and the definition (11.3), we deduces

$$\mathbf{T}^{\perp}(\mathbf{k}) = \frac{4}{\sqrt{3}}\left[\frac{Z}{a}\right]^4\frac{[1+n^2]^{1/2}}{n[1-e^{-2n\pi}]^{1/2}}I(2+in,4,3;k)\sin\theta_0\mathbf{I} \qquad (11.20)$$

from wich we deduce (see [16], (11.60)–(11.64))

$$\mathbf{T}^{\perp}(\mathbf{k}) = \frac{\sqrt{3}e^{-2n\Theta}(2n^2K\cos n\phi - (1+n^2+n^2K^2)\sin n\phi)}{n[1+n^2]^{1/2}[1-e^{-2n\pi}]^{1/2}K^3}\sin\theta_0\mathbf{I}, \quad (11.21)$$

where

$$K = \frac{ka}{Z}, \quad \Theta = \frac{1}{2}\left(\cot^{-1}\left[\frac{n}{nK+1}\right] - \cot^{-1}\left[\frac{n}{nK-1}\right]\right)$$

$$\phi = \frac{1}{2}\ln\left[\frac{n^2+(nK+1)^2}{n^2+(nK-1)^2}\right].$$

To obtain the matrix element of the photoeffect, we have to give to k the value $k = (E_2 - E_1)/\hbar c$, with $E_2 - E_1 = E_2 - mc^2 + mc^2 - E_1$, where $mc^2 - E_1$ is the ionisation energy $Z^2\alpha^2mc^2/2$, and so, using (10.30), to write

$$\frac{E_2 - E_1}{\hbar c} = \left[\frac{1+n^2}{n^2}\right]\frac{Z^2\alpha^2}{2\alpha a}, \quad K = \left[\frac{1+n^2}{n^2}\right]\frac{Z\alpha}{2} \qquad (11.22)$$

We see that for values of Z not too large, the retardation is not very important in the photoeffect, at least in the non relativistic calculation.

11.5 A recapitulative Verification

1. We can obtain a verification of (11.22), in which $\theta_0 = \pi/2$ and so $\mathbf{I} = \mathbf{e}_3$, by the passage to the matrix element used in the calculation of the photoeffect without retardation.

Using the relation the for small values of K

$$2n^2K\cos n\phi - (1+n^2+n^2K^2)\sin n\phi \simeq \frac{8n^4K^3}{3(1+n^2)}, \qquad (11.23)$$

which may be proved by means of Taylor developments, we obtain the modulus of the vector $\mathbf{U}$ of the transition in the Schrödinger approximation, that we will denote $\mathbf{U}_a$ as in Sect. 7.2,

$$|\mathbf{U}_a| = \lim_{k \to 0} |2\mathbf{T}(\mathbf{k})| = U_a(n), \quad U_a(n) = 2 \, \frac{8n^3 e^{-2n \cot^{-1} n}}{\sqrt{3}[1 + n^2]^{3/2}[1 - e^{-2n\pi}]^{1/2}} \tag{11.24}$$

then using

$$\frac{\mathbf{U}_a}{2} = \frac{\hbar}{mc} \int \psi_2 \nabla \psi_1 d\mathbf{r} = \frac{E_1 - E_2}{\hbar c} \int \psi_2 \mathbf{r} \psi_1 d\mathbf{r}, \quad |X_{21}| = \left| \int \psi_n \mathbf{r} \psi_1 d\mathbf{r} \right| \tag{11.25}$$

(see [5], Eq. (59.4)) we obtain, with the value (11.22) of $E_2 - E_1$), the well-known expression ([5], eq.(71.4))

$$|X_{21}|^2 = \frac{2^8 e^{-4n \cot^{-1} n}}{3(1 - e^{-2n\pi})} \left[\frac{n^2}{1 + n^2} \right]^5 \frac{(\alpha a)^2}{(Z\alpha)^4} \tag{11.26}$$

used in the calculation of the photoeffect without retardation.

2. On the other hand, using the approximation $\varepsilon + 1 = 2$, $\nu = n$, equivalent to the Pauli approximation of the Dirac radial functions, we see immediatly on (11.8) and (11.13) that, denoting $\mathbf{U}_b$ and $\mathbf{U}_c$ the vector $\mathbf{U}_2$ corresponding to the transitions $s1/2 - p1/2$ and $s1/2 - p3/2$ respectively, we can write

$$|\mathbf{U}_b| = \frac{1}{\sqrt{3}} U_a(n), \quad |\mathbf{U}_c| = \frac{\sqrt{2}}{\sqrt{3}} U_a(n) \tag{11.27}$$

in conformity with the relation (7.7), which holds in the Pauli approximation for all transitions $s1/2 - p1/2$ and $s1/2 - p3/2$, now including a state p of the continuum

$$\mathbf{U}_a^2 = \frac{\mathbf{U}_b^2}{3}, \quad \mathbf{U}_a^2 = \frac{2\mathbf{U}_c^2}{3}, \quad \mathbf{U}_a^2 = \mathbf{U}_b^2 + \mathbf{U}_c^2 \tag{11.28}$$

We can deduce from relations established in Sect. 9.2 that a direct passage of the vectors $\mathbf{T}^\perp(\mathbf{k})$ of the transitions $s1/2 - p1/2$ and $s1/2 - p1/2$ to a vector $\mathbf{T}^\perp(\mathbf{k})$ of a transition $s - p$ is not possible. In other words, one of the effect of the retardation is *to break the possibility to find an equivalence between the Pauli approximation and the Schrödinger theory*, and the reason lies on the incidence of the retardation on *the spherical parts* of the Dirac wave functions, related to the presence of the spin. The incidence is already sensible, in the transitions of the discrete spectrum (see (9.38), (9.39), (9.40)) and this incidence may be amplified in the contribution of the continuum, independently of the incidence of the chosen values for the radial functions.

Note : Integral formula implying the hypergeometric series.

The hypergeometric function, denoted $F(A, B, C, z)$, is defined by the series

$$F(A, B, C, z) = 1 + \frac{AB}{C} \cdot \frac{z}{1!} + \frac{A(A+1)B(B+1)}{C(C+1)} \cdot \frac{z^2}{2!} + .. \tag{11.29}$$

which:

- is reduced to a polynomial of degree p in z if A or B is a negative integer $-p$,
- if not, is an holomorphic function defined on the disk of convergence $|z| < 1$ of the series.

In both case it allows the calculation of integrals by means of the following formula

$$\int_0^\infty e^{-\lambda \rho} \rho^{\beta-1} F(A, C, \mu\rho) \mathrm{d}\rho = \frac{\Gamma(\beta)}{\lambda^\beta} F(A, \beta, C, \frac{\mu}{\lambda}) \qquad (11.30)$$

with, in the case (b), $Re(\beta) > 0, Re(\lambda) > Re(\mu) > 0$ (see [45], p.278).

This formula is nothing else but the integration term by term of the series

$$e^{-\lambda \rho} \rho^{\beta-1} (1 + \frac{A}{C} \cdot \frac{\mu\rho}{1!} + \frac{A(A+1)}{C(C+1)} \cdot \frac{(\mu\rho)^2}{2!} + ..)$$

with the help of the relations ([45], p.9)

$$\int_0^\infty e^{-\lambda \rho} \rho^{\beta+p-1} \mathrm{d}\rho = \frac{\Gamma(\beta+p)}{\lambda^{\beta+p}}, \quad \Gamma(\beta+p) = \Gamma(\beta)\beta(\beta+1)..(\beta+p-1)$$

The following formulas are usefull :

$$F(A, B, C, z) = (1-z)^{C-A-B} F(C-A, C-B, C, z) \qquad (11.31)$$

$$F(A, B, C, z) = \Gamma(C) [\frac{\Gamma(C-A-B)}{\Gamma(C-B)\Gamma(C-A)} F(A, B, A+B-C+1, 1-z)$$

$$+(1-z)^{C-A-B} \frac{\Gamma(A+B-C)}{\Gamma(A)\Gamma(B)} F(C-A, C-B, C-A-B+1, z)]$$

$$(11.32)$$

(see [45], p. 47).

Matrix Elements for the Relativistic Transitions with Retardation 1S1/2-Continuum

Abstract. This chapter gives exact calculation of the matrix elements of the transition by the use of hypergeometric series and its verification by using the Laplace method. The incidence of the diverse approximations with regard to the exact solution are drawn out numerically. A conclusion is the necessity of the use of retardation and considerable divergence between relativistic and nonrelativistic approachs for the high values of energy in the continuum.

12.1 General Formulas

The formulas established here allow us to calculate the matrix elements, up to all wanted degree of precision. They concern any Z number, provided that the potential created by the nucleus can be supposed of the form Ze/r. For large Z, a high degree of precision would not be compatible with the absence of corrections due to the size of the nucleus. However, concerning the transitions to the continuum, these corrections are probably negligible.

For $Z \geq 2$ a high degree of precision would require that the atom is considered as *strictly* hydrogen-like. Precise results cannot be obtained, for example, with the usual screening correction approximations to the value of Z, which is made for the K shell. Nevertheless, experiments are made with atoms whose all electrons except one have been drived away, and for these experiments, except the question of the size of the nucleus, the calculation is to be considered as suitable as for the hydrogen atom.

We will only consider here the transitions 1S1/2-continuum and the cases where the variation in the transition of the magnetic number m is null.

The calculation for the others states of the discrete spectrum and for the cases where the variation of m is equal to ± 1 can be made on the same model.

We recall the formulas we need.

$$L_{s,ij}(k) = \int_0^\infty j_s(kr)g_i(r)f_j(r)r^2\mathrm{d}r, \qquad (12.1)$$

where $j_s(x)$ represents the spherical Bessel function such that $s = 0$ or $s = 2$ and $g_i(r)$, $f_i(r)$ ($i = 1, 2$) are the radial components of the Darwin solutions of the states ψ_1, ψ_2.

1. *Transitions $S1/2 - P1/2$: $m_1 = m_2 = 0$, ($\varepsilon = 1$) and $m_1 = m_2 = -1$, ($\varepsilon = -1$).*

$$\mathbf{T}_1^\perp(\pm\mathbf{k}) = 0 \tag{12.2}$$

$$\mathbf{T}_2^\perp(\pm\mathbf{k}) = \frac{\varepsilon}{3}[(3L_{0,12} + L_{0,21} - 2L_{2,21})(k)]\sin\theta_0\ \mathbf{I} \tag{12.3}$$

2. *Transitions $S1/2 - P3/2$: $m_1 = m_2 = 0$, ($\varepsilon = 1$) and $m_1 = m_2 = -1$, ($\varepsilon = -1$).*

$$\mathbf{T}_1^\perp(\pm\mathbf{k}) = -\varepsilon\frac{3}{\sqrt{2}}[(L_{2,12} + L_{2,21})(k)]\cos\theta_0\sin\theta_0\ \mathbf{J} \tag{12.4}$$

$$\mathbf{T}_2^\perp(\pm\mathbf{k}) = \frac{\sqrt{2}}{6}[(4L_{0,21} - 3L_{2,12} + L_{2,21})(k)]\sin\theta_0\ \mathbf{I} \tag{12.5}$$

(a) The radial functions.

The radial functions g_1, f_1 corresponding to the state $\psi_1 = 1S1/2$ ((17.14) will be written here

$$g_1(r) = C\left[\frac{Z}{a}\right]^{3/2}2e^{-\rho}\rho^{\gamma_1-1}, \quad f_1(r) = -Cd\alpha Z\left[\frac{Z}{a}\right]^{3/2}e^{-\rho}\rho^{\gamma_1-1} \tag{12.6}$$

with

$$\gamma_1 = [1 - \alpha^2 Z^2]^{1/2}, \quad \delta = \left[1 + \frac{\alpha^2 Z^2}{\gamma_1^2}\right]^{-1/2}$$

$$C = 2^{\gamma_1-1}\left[\frac{1+\delta}{\Gamma(2\gamma_1+1)}\right]^{1/2}, \quad d = \frac{2}{\alpha Z}\left[\frac{1-\delta}{1+\delta}\right]^{1/2}$$

and

$$\rho = \frac{Zr}{a} \tag{12.7}$$

Let us denote

$$\gamma = [\kappa^2 - \alpha^2 Z^2]^{1/2}, \quad \varepsilon = \frac{E}{mc^2}, \quad n = \frac{\alpha Z}{[\varepsilon^2 - 1]^{1/2}}, \quad \nu = [n^2 + \alpha^2 Z^2]^{1/2} \tag{12.8}$$

the parameters that are associated with the state ψ_2.

Then the radial functions of a state ψ_2 (energy E) in the continuum, (10.10), is written

$$\left\{\begin{array}{c}g(r)\\f(r)\end{array}\right\} = [\varepsilon\pm1]^{1/2}\left[\frac{Z}{a}\right]^{3/2}\frac{e^{\nu\pi/2}|\Gamma(\gamma+1-i\nu)|\sqrt{n}}{\alpha Z\sqrt{\pi}\ \Gamma(2\gamma+1)}\left[\frac{2}{n}\right]^\gamma\rho^{\gamma-1}\left\{\begin{array}{c}\text{Re}\\\text{Im}\end{array}\right\}S$$

$$S = \left[\frac{-\kappa - in}{\gamma + i\nu}\right]^{1/2} e^{-i\rho/n} F(\gamma + i\nu; 2\gamma + 1; i\frac{2\rho}{n}) \tag{12.9}$$

The allowed transitions are such that ψ_2 is a state $P1/2$ ($\kappa = 1$) or a state $P3/2$ ($\kappa = -2$).

(b) The integrals $L_{s,ij}$

Let us define

$$K = \frac{ka}{Z} \tag{12.10}$$

The determination of the integrals (12.1) implies the calculation of integrals in the form

$$I_s(K, n) = \int_0^\infty \exp[-(1 + \frac{i}{n})\rho]j_s(K\rho)\rho^{\gamma_1 + \gamma}F(\gamma + i\nu; 2\gamma + 1; i\frac{2\rho}{n})d\rho, \tag{12.11}$$

where $s = 0, 2$.

Using the equalities

$$[\varepsilon + 1]^{1/2} = \frac{[\nu + n]^{1/2}}{\sqrt{n}}, \quad [\varepsilon - 1]^{1/2} = \frac{\alpha Z}{\sqrt{n}[\nu + n]^{1/2}} \tag{12.12}$$

and denoting

$$H(n) = C\frac{e^{\nu\pi/2}|\Gamma(\gamma + 1 - i\nu)|\sqrt{n}}{\sqrt{\pi}\,\Gamma(2\gamma + 1)}\left[\frac{2}{n}\right]^\gamma\left[\frac{-\kappa - in}{\gamma + i\nu}\right]^{1/2} \tag{12.13}$$

we can write

$$L_{s,12}(k) = \frac{2}{\sqrt{n}[\nu + n]^{1/2}}\text{Im}[H(n)I_s(k, n)] \tag{12.14}$$

$$L_{s,21}(k) = -d\frac{[\nu + n]^{1/2}}{\sqrt{n}}\text{Re}[H(n)I_s(k, n)] \tag{12.15}$$

The substitution of these real numbers in the (12.2)–(12.5) (and (9.20)–(9.29) for all the degenerencies) gives the values of the vectors $\mathbf{T}_j^\perp(\pm\mathbf{k})$.

Note that for verifying the calculations which use the dipole approximation, we have to consider also the integral $I_0(0, n)$, in which $j_0(K\rho)$ is replaced by unity.

12.2 Numerical Calculation of the Formulas

The calculation of $H(n)$ does not present difficulties, but the one of integrals in the form

$$I_s(K, n) = \int_0^\infty e^{-\lambda\rho}\rho^p j_s(K\rho)F(a; c; \mu\rho)d\rho \tag{12.16}$$

with

$$j_0(x) = \frac{\sin x}{x}, \quad j_2(x) = 3\left[\frac{\sin x}{x^3} - \frac{\cos x}{x^2}\right] - \frac{\sin x}{x}, \tag{12.17}$$

where $\lambda = 1 + i/n$, $a = \gamma + i\nu$, $p = \gamma_1 + \gamma$, $c = 2\gamma + 1$, $\mu = i2/n$, requires explanations.

These integrals imply confluent hypergeometric functions and two ways may be envisaged for their calculation.

(a) Analytic calculation.

For a long time it has been the only way, by the use of the Laplace method. Such a method is based on the representation in the complex plane of the confluent hypergeometric functions and the use the residues theorem. But it needs the approximation $Z^2\alpha^2 \ll \kappa^2$, which allows one to replace in (12.16) the numbers a, c, p by integers.

This method is suitable only for small values of Z, but allows a verification of the validity of the results obtained by the second way below. We recall that it is the natural continuation of an approach initiated in [55] and which has been used in the study of the photoeffect especially in [5, 32], Sect. 71.

We will not detail the calculation made in [18] and will only recall the numerical results given by (12.41)–(12.48) of this article.

(b) Numerical calculation on computers.

The calculation of the integrals $L_{s,ij}$, (12.1), may be presently achieved by means of sophisticated computer softwares.

Nevertheless, given the good convergence of the hypergeometric series, a simple calculation can be made with the use of (11.31) (except for the calculation of $L_{2,21}$, in the case of P1/2, which implies more elaborated processes, but this integral is negligible for small values of the energy in the continuum).

12.3 Some Numerical Results

The numerical results presented here concern the square of the matrix elements deduced from (12.2), (12.3) for $P1/2$ and (12.4), (12.5) for $P3/2$, in which $\theta_0 = \pi/2$, i.e. the number

$$L(k) = |\mathbf{T}_2^{\perp}(\mathbf{k})|^2, \quad \text{where} \quad k = |\mathbf{k}| = \frac{E - E_1}{\hbar c} = \frac{2\pi}{\lambda}, \quad \mathbf{k}.\mathbf{e}_3 = 0 \tag{12.18}$$

E_1 is the energy of $1S1/2$ and E the energy of a state $P1/2$ or $P3/2$ of the continuum, λ the wave length of the incident wave.

The number $L(k)$ is dimensionless but to be in conformity with the definition of [5], Eq. (73.1), it may be considered as a quantity expressed in unit $(mc^2/\hbar)^2 = (1/\alpha a)^2$.

The calculation gives results obtained by the Laplace method and their comparison with the ones obtained by the use of the hypergeometric series.

In the limit $k \rightarrow 0$, (12.41)–(12.48) of [18] allow one to recover the expressions of the numbers $L(0) = |\mathbf{T}_2^{\perp}(0)|^2 = |\mathbf{U}_2|^2/4$ (see (12.8), (11.13)) corresponding to the relativistic, non retarded case.

In the Pauli approximation (here $\nu = n$ in addition to $Z^2\alpha^2 \ll \kappa^2$) the numbers $L(0)$ satisfy the relations $L(0) = |X_{E1}|^2/3$ and $L(0) = 2|X_{E1}|^2/3$ for the transitions s1/2-p1/2 and s1/2-p3/2, respectively (see 11.29), where $|X_{E1}|^2$ (see [5], Eq. (71.4)) is the matrix element of a transition $1s - p$ in the Schrödinger theory. (We recall that this matrix element has been used for a long time as the main element of the theoretical verification of the experiments on the photoeffect ([5], Sect. 71)). This property has been analytically verified on the relativistic non retarded formulas of the Sect. 12.10, and numerically verified for the small values of k on the relativistic retarded formulas of Sect. 11.4.

Here we compare $L(k)$ with the non relativistic retarded matrix elements of the transitions $1s - p$ established in conformity with the calculation of [55] (see (11.22) with $\theta_0 = \pi/2$) (multiplied by 1/3 and 2/3 for the transitions 1s1/2-p1/2 and 1s1/2-p3/2, respectively). That gives the error made by the use of the nonrelativistic retarded theory.

1. In Tables I, II, III, the line (a) gives the values $L_e(k)$ of $L(k)$ calculated with the use of hypergeometric series, and which are obtained by means of (11.31)

We emphasize that the high degree of precision allowed by this method concerns only atoms considered strictly as hydrogenic and whose number Z in not too large, in such a way that the corrections due to the presence of other electrons or the size of the nucleus are not to be taken into account.

For simplicity, because the calculation of $I_2(K, n)$ for the transitions 1S1/2-P1/2 requires sophisticated numerical methods, we have leaved out the calculation of this number. For the values of $K = ak/Z$ not too large its contribution is negligible. For the large values of K, the matrix elements of these transitions are not mentioned.

Table 12.1 is relative to the numbers

$L_e(k)$: hypergeometric series.

$L_a(k)$: Laplace method.

$L(0)$, $(k = 0)$: relativistic non retarded (see (11.8) and (11.9)).

$L_0(k)$: non relativistic (Schrödinger) retarded (see (11.22)) multiplied by 1/3 and 2/3 for the transitions $1s1/2 - p1/2$ and $1s1/2 - p3/2$, respectively.

The line (b) gives the ratios $L_a(k)/L_e(k)$ and allows a numerical comparison between the two methods.

The line (c) gives the ratios $L(0)/L_e(k)$ and allows the evaluation of the incidence of the retardation in the relativistic calculation.

The line (d) gives the ratios $L_0(k)/L_e(k)$. It allows the estimation of the errors which are made when the retarded non relativistic calculation is used.

Table 12.1. Matrix elements $L(k)$ (see [19])

$\lambda(\mathring{A})$	$E - mc^2$ (Kev)	$K = ak/Z$		$1S1/2 - P1/2$	$1S1/2 - P3/2$
		I:H($Z = 1$)			
828.42	0.00136	0.004	a	0.110973	0.221954
			b	1.00005	1.00007
			c	1.00004	1.00003
			d	1.00005	1.00001
0.608	20.4	5.47	a	1.0997×10^{-8}	2.3179×10^{-8}
			b	0.99988	0.99982
			c	0.9357	1.0034
			d	1.1080	1.0514
0.0091	1360.6	364.9	a	3.37×10^{-13}	6.035×10^{-13}
			b	0.99963	1.0021
			c	0.1479	0.3276
			d	0.9434	1.0553
0.006	2040.8	547.3	a	–	3.032×10^{-13}
			b	–	1.0024
			c	–	0.1625
			d	–	0.093
9×10^{-5}	136056	36487	a	–	9.40×10^{-16}
			b	–	1.001
			c	–	4.83×10^{-6}
			d	–	2.44×10^{-11}
		II:Na($Z = 11$)			
7.457	0.0165	0.041	a	0.12723	0.25549
			b	1.006	1.009
			c	1.005	1.003
			d	1.006	1.002
0.502	23.05	0.602	a	5.692×10^{-4}	1.2282×10^{-3}
			b	1.0004	0.995
			c	0.927	1.006
			d	1.141	1.057
0.075	162	4.01	a	3.73×10^{-6}	1.3317×10^{-5}
			c	0.585	0.997
			d	2.6	1.45
0.015	821	20	a	–	2.283×10^{-7}
			c	–	0.611
			d	–	4.8

(Continued)

Table 12.1. *Continued*

$\lambda(\mathring{A})$	$E - mc^2$ (Kev)	$K = ak/Z$		$1S1/2 - P1/2$	$1S1/2 - P3/2$
		III:Cs($Z = 55$)			
0.3	0.041	0.2	a	0.10806	0.24347
			b	1.170	1.251
			c	1.139	1.070
			d	1.164	1.033
0.2	20.57	0.301	a	0.05	0.1226
			c	1.067	1.071
			d	1.301	1.038
0.03	370	2	a	–	0.00219
			c	–	0.955
			d	–	1.8

$$L(k) = |\mathbf{T}_2^{\perp}(\mathbf{k})|^2, \quad k = |\mathbf{k}| = \frac{E - E_1}{\hbar c} = \frac{2\pi}{\lambda}, \quad \mathbf{k}.\mathbf{e}_3 = 0$$

$L_e(k)$ (hypergeometric series), $L_a(k)$ (Laplace method), $L(0)$ relativistic non retarded, $L_0(k)$, non relativistic retarded.
$a = L_e(k), b = L_a(k)/L_e(k), c = L(0)/L_e(k), d = L_0(k)/L_e(k)$.

Comments.

For E close to mc^2, i.e. for K small, and small values of Z, all the lines (b) to (d) must be close to 1, as can be seen on Table I ($Z = 1$) for K=0.004. That constitutes a very credible confirmation of the validity of all the formulas used.

For small values of Z the line (b) must be close to 1 as that can be seen on table I (Z=1) and II ($Z = 11$). So, for these values, the formulas (12.27)–(12.34) whose running time on a computer is shorter than for the ones of the hypergeometric series, can be used. But for large values of Z, the table III shows that the Laplace method gives wrong results.

The incidence of the retardation, in the relativistic calculation, begins to be important (for $Z = 1$) around $E - mc^2 = 300$ Kev and then becomes very large, as it is schown in the line (c).

The errors (line (d)) due to the use of the retarded non relativistic formulas, with respect to the retarded relativistic ones, are not very large for $E - mc^2 < 1,300$ Kev ($Z = 1$), but then they increase fastly, and become considerable for E very large. That shows the profound difference between the relativistic and the nonrelativistic calculations. One can notice on Table 12.1 incidence of the value of Z on these errors.

We recall that the nonrelativistic calculation cannot be applied in the cases of degeneracies (for reasons analog to the differences between the normal and anomal Zeeman effect), as it is schown by the formulas (9.20)–(9.29).

12.4 Conclusion

We have a numerical confirmation of the validity of the formulas giving the exact relativistic matrix elements with retardation of the photoeffect of hydrogenic atoms, by the good concordance, for small walues of Z, of the results in the two ways of calculation which have been employed, the exact one, based on the use of the hypergeomtric series, and the Laplace method which implies the approximation $\alpha^2 Z^2 \ll \kappa^2$.

An important point is the necessity of the use, in the photeffect, of the relativistic calculation in place of the nonrelativistic one, even when the retardation is taken into account in this last calculation.

13

The Radiative Recombination

Abstract. This chapter concerns what is called the radiative recombination, that is, the inverse of the photoeffect: one considers the emission of a plane wave after the capture by a bare nucleus of an electron whose state is placed in the continuum.

13.1 Motivations and Definition of Cross Sections

The radiative recombination (RR) for an hydrogenic atom is the inverse of the photoeffect. Instead of considering that a photon of energy $h\nu = \hbar\omega$ falls on the electron, bound in a state of energy E_1, and that the electron jumps to a state of the continuum of energy $E_2 = E_1 + \hbar\omega$, one supposes that an electron in a state of the continuum of energy E_2 may be captured by the bare nucleus until a bound state of energy E_1 with the simultaneous emission of a photon whose energy is $\hbar\omega$.

The recent studies about this process has been achieved in particular by Jörg Eichler and Akira Ichihara (see [24] and [38], named here [I/E]). Such a process *"plays an important role in plasma physics, in particular for the spectroscopic analysis of fusion plasmas"* ([I/E], p. 2).

Given the kinetic energy T of the incident electron, so that ([E/I], (13.7))

$$T = E_2 - mc^2 = \hbar\omega - \epsilon \quad (E_1 = mc^2 - \epsilon, \ \epsilon > 0) \tag{13.1}$$

cross sections σ_{RR} for the study of RR are considered. We will follow the definition given in [I/E], (13.11)

$$\sigma_{RR} = \frac{\sigma_{ph}}{f} \tag{13.2}$$

(a) σ_{ph} is defined as ([I/E], (13.6))

$$\sigma_{ph} = \frac{8\pi^2 \alpha mc^2 (\lambda_C^R)^2}{\hbar\omega(2j+1)} L, \tag{13.3}$$

where $j + 1/2 = |\kappa|$ and κ is the principal quantum number of the state of energy E_1, and $\lambda_C^R = \hbar/mc$ is the reduced Compton wavelength.

The number L is defined as

$$L = \frac{1}{4\pi} \int_0^{2\pi} \int_0^\pi ([\mathbf{T}_1^\perp(\mathbf{k})]^2 + [\mathbf{T}_2^\perp(\mathbf{k})]^2) \sin\theta_0 d\varphi_0 d\theta_0 \qquad (13.4)$$

The vectors $\mathbf{T}_j^\perp(\mathbf{k})$ allow one to define the matrix elements, (8.1), (8.3), of the transition from the state of energy E_2 to the state of energy E_1. The propagation vector $\mathbf{k}$ of the emitted wave is such that $|\mathbf{k}| = k = \hbar\omega$ and θ_0, such that $\mathbf{K}.\mathbf{e}_3 = \cos\theta_0$, $(\mathbf{K} = \mathbf{k}/k)$, defines the angle between $\mathbf{k}$ and the $\mathbf{e}_3$ direction of the Darwin solutions of the Dirac equation. So (8.4) corresponds to an average upon all the directions of the vector $\mathbf{k}$.

Note that σ_{ph} may be also written

$$\sigma_{ph} = 2\pi\alpha^2 \frac{a\lambda}{|\kappa|} L, \quad (\lambda = \frac{2\pi\hbar c}{k}), \qquad (13.5)$$

where $a = \hbar/\alpha mc$ is the Bohr radius and λ the wavelength of the emitted photon.

If a and λ are both expressed in Angström ($10^{-8}cm$) and σ_{ph} in barn ($10^{-24}\,cm^2$), it is necessary to multiply the right hand part of (13.5) by a conversion factor equal to $(10^{-8})^2 \times 10^{24} = 10^8$.

(b) The number f is defined ([I/E], Eq. (12)) as

$$f = \frac{\tilde{T}^2 + 2\tilde{T}}{(\tilde{T} + \tilde{\epsilon})^2}, \quad \tilde{T} = \frac{T}{mc^2}, \quad \tilde{\epsilon} = \frac{\epsilon}{mc^2}, \quad mc^2 = 510.99906\,\text{keV} \qquad (13.6)$$

in which $\tilde{T}$, $\tilde{\epsilon}$ and mc^2 are expressed in Kiloelectronvolt.

13.2 Some Numerical Results

The values of σ_{RR} have been published in [I/E], p. 10-121, for the bound states 1S1/2, 2S1/2,...3D5/2, from $Z = 1$ ($1 \leq T \leq 5 \times 10^4$ eV) to $Z = 112$ ($1 \leq T \leq 8 \times 10^6$ eV).

We give here the way of calculating these values for 1S1/2 by means of the formulas established in the previous sections and gathered together in the Chapter 11.

The number L will be considered as corresponding to the transitions P1/2-1S1/2 and P3/2-1S1/2, in the case where the variation in the transition of the magnetic number is null (linear polarization). Furthermore, L will be considered in (13.4) as corresponding to the sum of these two transitions.

The radial functions g_1, f_1 of the state of energy E_1 are given by (12.6) where ϵ is given by (4.53) with $n = 1$, $|\kappa| = 1$:

$$\epsilon = mc^2 \frac{Z^2\alpha^2}{2}(1 + \frac{Z^2\alpha^2}{4}) \qquad (13.7)$$

The radial functions g_2, f_2 for P1/2 and P3/2 are given by (12.9) ($|\kappa| = 1$ and $|\kappa| = 2$) by means of the suitable value of the number n of (12.8) corresponding to the energy E_2.

The value of the component of L for P1/2 is given by (12.2) and the (12.3) with an integrating factor $8\pi/3$, for P3/2 by (12.4) with a factor $8\pi/15$ and (12.5) with a factor $8\pi/3$.

The values of σ_{RR} (expressed in barn) in function of T (expressed in eV) are rounded off at three numerals in [I/E]. We give some values, rounded off at four numerals, obtained by the way of calculation based on the use of hypergeometric series (Sect. 12.2) and verified (for $Z = 1, 2$ only, for larger values of Z the verification is less legitimated) by the Laplace method:

$Z = 1$: $1\,eV$: $1088b$, $10\,eV$: $75.93b$, $100\,eV$: $1.652b$
$Z = 2$: $1\,eV$: $4512b$, $10\,eV$: $406.6b$, $100\,eV$: $19.80b$
$Z = 11$: $1\,eV$: $1.380 \times 10^5 b$, $10\,eV$: $1.350 \times 10^4 b$

These values are in agreement with those obtained by Professor Ichihara (private communication, 2001).

Part V

Interaction with a Magnetic Field

14

The Zeeman Effect

Abstract. This chapter is relative to the calculation in other external fields, limited here to a weak magnetic field, giving one of the most important phenomena associated with the Dirac theory, the anomalous Zeeman effect.

14.1 An Approximation Method for Time-Independent Perturbation

We consider the electron of an hydrogenic atom that is submitted to a potential $A \in M$ in the form

$$A = A^0 e_0 + \mathbf{A} e_0, \quad eA^0 = V(r) = \frac{e^2 Z}{r} \tag{14.1}$$

We suppose that $\mathbf{A}$ is sufficiently small for considering its incidence as a perturbation of the Darwin solution corresponding to a state, in the central potential A_0, of energy E. The energy E' of the electron will be then written in the form

$$E' = E + \Delta E \tag{14.2}$$

The method of perturbation that we are going to use is based on the following hypothesis.

1. The wave fonction ψ may be considered in the form

$$\psi(x^0, \mathbf{r}) = \phi(\mathbf{r}) e^{-\mathbf{i} e_3 (E'/\hbar c) x^0} \tag{14.3}$$

in such a way that (4.6) becomes

$$\nabla \phi = \frac{1}{\hbar c} [-E_0 \bar{\phi} + (E + \Delta E + V - e\mathbf{A})\phi] \mathbf{i} e_3, \quad E_0 = mc^2, \quad \bar{\phi} = e_0 \phi e_0 \tag{14.4}$$

2. Both $\phi(\mathbf{r})$ corresponds to the Darwin solution for the state whose energy is E, and (14.4) is acceptable in average by means of an integration on the E^3 space of a formula *in which the Dirac current*

$$j = \phi e_0 \tilde{\phi} = j^0 e_0 + \mathbf{j} e_0$$

of the state intervenes (H. Krüger, 1991, private communication).

Multiplying (14.4) on the right by $e_0\underline{i}e_3\tilde{\phi}e_0$, taking into account (4.6) we can write

$$\Delta E \int \phi e_0 \tilde{\phi} e_0 \, d\tau = e \int \mathbf{A} \phi e_0 \tilde{\phi} e_0 d\tau \qquad (14.5)$$

Applying (4.18))

$$\int \phi e_0 \tilde{\phi} e_0 \, d\tau = \int j e_0 \, d\tau = \int (j^0 + \mathbf{j}) \, d\tau = 1$$

we obtain

$$\Delta E = e \int \mathbf{A}(j^0 + \mathbf{j}) \, d\tau = e \int (j^0 \mathbf{A} + \mathbf{A}.\mathbf{j} + \mathbf{A} \wedge \mathbf{j}) \, d\tau. \qquad (14.6)$$

Since ΔE is a scalar, the right-hand part of this equation must be a scalar and so the two following relations must be verified

$$\int j^0 \mathbf{A} \, d\tau = 0 \qquad (14.7)$$

$$\int \mathbf{A} \wedge \mathbf{j} \, d\tau = 0 \qquad (14.8)$$

Then we can write

$$\Delta E = e \int \mathbf{A}.\mathbf{j} \, d\tau. \qquad (14.9)$$

14.2 The Margenau Formula: The Landé Factor

We consider that the atom is submitted to a magnetic field in the form $\mathbf{H} = H\mathbf{e}_3$ where H is constant. The corresponding potential $\mathbf{A}$ is such that we can write

$$\mathbf{A} = \frac{H}{2}(\mathbf{e}_3 \times \mathbf{r}) \quad \Rightarrow \quad \mathbf{H} = \nabla \times \mathbf{A} = H\mathbf{e}_3 \qquad (14.10)$$

Indeed,

$$\mathbf{A} = \frac{H}{2} r \sin \theta \mathbf{v} \qquad (14.11)$$

and, using (4.1), it is easy to verify that

$$\frac{H}{2} \nabla \times (r \sin \theta \mathbf{v}) = H\mathbf{e}_3$$

Since j^0 is independent of φ (4.17a) and $\int_0^{2\pi} \mathbf{v} d\varphi = 0$, (14.7) is verified. Since $\mathbf{j}$ is colinear to $\mathbf{v}$ (4.17b), (14.8) is also verified.

So if $\mathbf{H}$ is sufficiently weak in such a way that its effect may be considered as a perturbation of the Darwin solution, (14.9) may be applied.

The shift ΔE of the state of energy E is called the Zeeman effect, more precisely "anomalous effect," with respect to the shift obtained in the nonrelativistic theory named "normal effect."

We deduce from (14.9), (14.11), (4.17b),

$$\Delta E = eH \times I \times J \tag{14.12}$$

$$I = 2\pi \int_0^\pi ((M^2 - L^2)\sin^2\theta + 2LM\cos\theta\sin\theta)\sin\theta\,d\theta \tag{14.13}$$

$$J = \int_0^\infty (gfr)r^2\,dr \tag{14.14}$$

(a) *Calculation of I.*
We will use the fact that L, M are in the form

$$L = \frac{C}{\sqrt{2\pi}}P_l^m, \quad M = \frac{D}{\sqrt{2\pi}}P_l^{m+1}$$

(4.30), (4.31) and the relation

$$\int_0^\pi P_j^r P_k^r \sin\theta\,d\theta = \delta_{jk}$$

Writting $\sin^2\theta = 1 - \cos^2\theta$ in (14.13) and $I = I_1 + I_2$,

$$I_1 = 2\pi \int_0^\pi (M^2 - L^2)\sin\theta\,d\theta = D^2 - C^2 \tag{14.15}$$

$$I_2 = 2\pi \int_0^\pi (-(M^2 - L^2)\cos^2\theta + 2LM\cos\theta\sin\theta)\sin\theta\,d\theta \tag{14.16}$$

using (6.3), (6.4) for the integration of the terms $(M\cos\theta)^2$, $(L\cos\theta)^2$, $(L\sin\theta)(M\cos\theta)$ in I_2, one obtains, after surprising simplifications in the calculation of I_2,

1. $\kappa = -(l+1)$:

$$I_1 = -\frac{2m+1}{2l+1}, \quad I_2 = \frac{2m+1}{(2l+1)(2l+3)}$$

$$I = -\frac{(2m+1)2(l+1)}{(2l+1)(2l+3)} = \frac{(2m+1)2\kappa}{(2\kappa+1)(2\kappa-1)} \tag{14.17}$$

2. $\kappa = l$

$$I_1 = \frac{2m+1}{2l+1}, \quad I_2 = \frac{2m+1}{(2l+1)(2l-1)}$$

$$I = \frac{(2m+1)2l}{(2l+1)(2l-1)} = \frac{(2m+1)2\kappa}{(2\kappa+1)(2\kappa-1)} \tag{14.18}$$

(b) *Calculation of J:*
Deducing fgr from (4.12) we can write

$$J = \frac{\hbar}{2m_e c} \int_0^\infty [g^2 + f^2 + \kappa(g^2 - f^2) + (gg' + ff')r]r^2 dr,$$

where m_e means the mass of the electron (for evoiding a confusion with $m \in Z$). Since

$$\int_0^\infty (g^2 + f^2)r^2 dr = 1 \Rightarrow \kappa \int_0^\infty (g^2 - f^2)r^2 dr = \kappa(1 - 2\int_0^\infty f^2 r^2 dr)$$

$$\int_0^\infty (gg' + ff')r^3 dr = \frac{1}{2}[(g^2 + f^2)r^2]_0^\infty - \frac{3}{2}\int_0^\infty (g^2 + f^2)r^3 dr = -\frac{3}{2}$$

and because $1 + \kappa - 3/2 = \kappa - 1/2$ we obtain

$$J = \frac{\hbar}{2m_e c}[\kappa - \frac{1}{2} - 2\kappa \int_0^\infty f^2 r^2 dr] \tag{14.19}$$

The Margenau formula:
Now we can deduce from (14.12)

$$\Delta E = H \frac{\hbar}{2m_e c}(2m + 1)\frac{2\kappa}{2\kappa + 1}[\frac{1}{2} - \frac{2\kappa}{2\kappa - 1}\int_0^\infty f^2 r^2 dr] \quad (m \in Z) \tag{14.20}$$

i.e. the Margenau formula [46], expressed here by means of the principal quantum number κ and the magnetic number $m \in Z$.

The consequence of this formula is that the levels of energy of the states corresponding to the different values of the magnetic number m appear as separated. So the number of the transitions between two states whose energies were E_1 and E_2 in the absence of a magnetic field is increased by new transitions between states of energies $E_1 + \Delta E_1(m_1)$ and $E_2 + \Delta E_2(m_2)$, where $\Delta E_1(m_1)$ and $\Delta E_2(m_2)$ are given by (14.20).

For a given walue of $\Delta m = m_1 - m_2$, the correspondent transitions are indescernible in an unperturbated experiment but, given the separation of the levels, there are as many observable transitions as different couple of numbers (m_1, m_2) when a magnetic field is present.

For the transitions $P1/2 - S1/2$ the couples (m_1, m_2) are

$$(0, 0), (-1, -1) \quad - . - \quad (0, -1) \quad - . - \quad (-1, 0)$$

For the transitions $P3/2 - S1/2$ the couples are

$$(0, 0), (-1, -1) \quad - . - \quad (1, 0), (0, -1) \quad - . - \quad (-1, 0), (-2, -1)$$

So two linear polarizations instead of one may be observed in the transitions $P1/2 - S1/2$ and $P3/2 - S1/2$, and four circular polarizations instead of two for the transitions $P3/2 - S1/2$,

The Landé Factor:

Introduced in the Pauli theory, the Landé factor G is defined as follows

$$G = \frac{j + \frac{1}{2}}{l + \frac{1}{2}}, \quad j + \frac{1}{2} = |\kappa| \tag{14.21}$$

(see [5], Eqs (46.6), (14.28)).

Denoting

$$\frac{\hbar}{2m_e c} = \mu_0, \quad m + \frac{1}{2} = m' \quad (m \in Z)$$

since

$$\kappa = -(l + 1), \quad l + 1 = j + \frac{1}{2} \Rightarrow \frac{2\kappa}{2\kappa + 1} = \frac{l + 1}{l + \frac{1}{2}} = \frac{j + \frac{1}{2}}{l + \frac{1}{2}}$$

$$\kappa = l, \quad l = j + \frac{1}{2} \Rightarrow \frac{2\kappa}{2\kappa + 1} = \frac{l}{l + \frac{1}{2}} = \frac{j + \frac{1}{2}}{l + \frac{1}{2}}$$

one can write as in [5], Eqs (47.1), in a form mixing the numbers j, l, κ but including the Landé factor,

$$\Delta E = H \mu_0 m' \left(\frac{j + \frac{1}{2}}{l + \frac{1}{2}} \right) [1 - \frac{\kappa}{\kappa - \frac{1}{2}} \int_0^\infty f^2 r^2 \mathrm{d}r] \tag{14.22}$$

Part VI

Addendum

The Contribution of the Discrete Spectrum to the Lamb Shift of the 1S1/2 State

Abstract. This chapter is devoted, in close relation with Part III, to the relativistic calculation of the contribution of the states of the discrete spectrum to the Lamb shift of 1S1/2.

15.1 The Lamb Shift

A complete description of one of most complex calculation concerning the hydrogenic atoms, the Lamb shift, is outside of the scope of our elementary presentation. However, the matrix elements of the transition between two states play an important role in this calculation, which cannot be omitted in our presentation.

A precise observation of the levels of energy of an electron in an hydrogenic atom shows a slight shift of the value of a level with respect to the one given by the Darwin solution of the Dirac equation. This phenomena has been observed for the first time by Lamb and Retherford [42] and is called the Lamb shift.

An interpretation of the shift of a level has been given by Bethe [4] as an incidence of all the virtual states, belonging to the discrete spectrum *and* the continuum, of the electron upon this level.

The calculation of the contribution of a state to the Lamb shift of a particular state is based on the consideration of three terms (see [25, 40]: the *Electrodynamics energy term* W_D, the *Electrostatic energy term* W_S, and the *Electromagnetic mass operator* W_M, in such a way that the contribution is in the form $W_D + W_S - W_M$.

For the contribution of the states whose energy is low, in particular those of the discrete spectrum, only the term W_D is taken into consideration. We will only consider this contribution.

The interest of the calculation of the contribution of the discrete spectrum is to refine the usual calculation in which, for the low energy contribution, an approximative formula is used.

Another interest lies in the fact that this contribution could have some importance in future experiments with atoms closed inside a cavity, in such a way that the contribution of the continuum would not to be taken into account (see [53] and [33]).

At least the study of the term W_D will allow us to show the difference of the values of this term between those obtained by the relativistic and nonrelativistic calculations. This difference is weak for the contribution of the discrete spectrum and the low levels of the continuum but becomes considerable (see Note below) for the ones of the continuum of high levels and explains the necessity of the mass renormalization represented by the term W_M.

We will consider only the case of the Lamb shift of the $1S1/2$ state of the hydrogen atom.

The formula giving the term W_D of the contribution to the shift of a state of energy E_2 to a state of energy E_1 is the following

$$\Delta E_{12} = \frac{\alpha(E_1 - E_2)}{4\pi^2} \int \frac{[\mathbf{T}_1^\perp(\mathbf{k})]^2 + [\mathbf{T}_2^\perp(\mathbf{k})]^2}{E_1 - E_2 - \hbar ck} \hbar c \, d\tau_0, \qquad (Ad.1)$$

where the vectors $\mathbf{T}_j^\perp(\mathbf{k})$ are defined by (8.3), $k = |\mathbf{k}|$, and, following the notations (φ, θ_0) of sec. 8,

$$\int f(\varphi, \theta_0, k) d\tau_0 = \int_0^{2\pi} \int_0^\pi \int_0^\infty f(\varphi, \theta_0, k) \sin \theta_0 d\varphi d\theta_0 dk.$$

It is to emphasize that this integral is convergent whatever the values of E_1 and E_2 may be.

15.2 Nonrelativistic Calculation

In the nonrelativistic calculation the Dirac equation is replaced by the Schrödinger one. The formula that is obtained (see [2]), which is convergent, is, if the dipole approximation is applied (i.e. $\mathbf{T}_j^\perp(\mathbf{k})$ are replaced by $\mathbf{T}_j^\perp(0)$), the formula used in [4] for the first calculation proposed for the explanation of the Lamb shift. But this last formula is divergent and its use implies that the integration upon k is cut off for a $k = kmax$. In [23] the value of $kmax = \alpha mc^2$ has been proposed and was used in the following calculations of the Lamb shift.

Here we are only interested in the calculation with retardation. A calculation of the contribution of the 2p state on the shift of 1s, in the case where $m_1 = m_2$ has been achieved in [7], Eq. (37), and gives

$$\frac{\Delta E_{12}}{\hbar c} = \frac{\alpha^4}{\alpha\pi} \cdot \frac{2^7}{3^8} [\ln \frac{4}{\alpha} - \frac{11}{12}] \quad \text{or } \nu = 85.6 \, \text{MHz} \qquad (Ad.2)$$

This formula is deduced from (7.4), (7.5) for the construction of the vectors $\mathbf{T}_j^\perp(\mathbf{k})$.

Exactly the same formula lies in [52], Eq. (16) but has been established by J. Seke quite independently.

For taking into account also the cases where $m_1 - m_2 = \pm 1$, and so the total contribution of the $2p$ states, this result is to be multiplied by 3 giving 256.8 MHz.

The total contribution of the p states of the discrete spectrum to the shift of 1s have been calculated in [8] and also quite independently in [31].

The values of the cumulative contributions due to the states 2, 3, 4...∞ are in these two articles :

Blaive. 2: 256.95; 3: 325.37; 4: 353.28: ...∞: 395.01. MHz

Seke.   2: 257.04; 3: 325.49; 4: 353.40; ...∞: 395.76. MHz

These values are not very different from the ones (see below) obtained in the relativistic calculation. But when the contribution of the continuum is taken into account especially for the levels of energies greater than αmc^2 the divergence is such that all comparison between the relativistic and the nonrelativistic calculation is to be abandoned (*see* Note below).

15.3 Relativistic Calculation

We have calculate in [14] the contribution of $2P1/2$ and $2P3/2$ to the shift of $1S1/2$. We have used (Ad.1) in the case where the magnetic numbers m_1, m_2 are null. For simplicity we have also used the Pauli-Schödinger approximation which gives results very close to the exact relativistic ones in such a way that $E_1 - E_2 = 3\alpha/8a$ and the expressions of $\mathbf{T}_j^{\perp}(\mathbf{k})$ (9.20), (9.21) and (9.24), (9.25), are reduced to Eqs (9.38), (9.39), (9.40). We obtain

$-2P1/2$:

$$\frac{\Delta E_{12}^1}{\hbar c} = \frac{\alpha^4}{a\pi} \cdot \frac{2^7}{3^8} \cdot \frac{1}{3}[\ln\frac{4}{\alpha} - \frac{137}{120} + \frac{6}{5}] \quad \text{or} \quad \nu = 33.7\,\text{MHz} \qquad (Ad.3)$$

$-2P3/2$:

$$\frac{\Delta E_{12}^2}{\hbar c} = \frac{\alpha^4}{a\pi} \cdot \frac{2^7}{3^8} \cdot \frac{2}{3}[\ln\frac{4}{\alpha} - \frac{137}{120} - \frac{3}{32}] \quad \text{or} \; \nu = 53.7\,\text{MHz} \qquad (Ad.4)$$

(note that the above number 137 is not to be confused with $1/\alpha$ but comes from $(1+1/2+1/3+1/4+1/5)/2=137/120$).

So the total contribution is $\nu = 33.7 + 53.7 = 87.34$ MHz close to the non relativistic 85.6.

The values obtained in [53] are about the same :

Table 9 : 101.073/3=33.69 MHz and Table 10 : 161.298/3=53.766 MHz.

In the same article J. Seke gives the values of the contributions of all the discrete spectrum (including those corresponding to the electric and magnetic quadrupole transitions) to the shift of $1S1/2$ and finds a total value of 429.7 MHz.

We recall his values only for the contributions of the $P1/2$ and $P3/2$ (corresponding to the electric dipole transitions) :

Table 9. P1/2. 2: 101.073; 3: 128.168; 4: 140.162; ... ∞ : 157.491. MHz

Table 10. P3/2. 2: 161.298; 3: 203.900; 4: 221.205; ... ∞ : 247.499. MHz

The sum of the contributions of the $P1/2$ and $P3/2$ is then

P1/2+P3/2. 2: 263.37; 3: 332.07; 4: 361.37; ... ∞ : 404.99. MHz

It to emphasize the good concordance of results obtained quite independently and despite the difference of the methods which have been used. On one side, the real formalism and the employ of the pure law of Maxwell without quantization. On the other, the spinors formalism and the use of the Quantum Field Theory.

15.4 Note

The experiments on the shift of $1S1/2$ give 8,173 MHz. A nonrelativistic calculation, implying not only the contribution of the discrete spectrum but also all the continuum, achieved by B. Blaive (see [17]) gives $396 + 4,759 = 5,146\,MHz$. One could think that the passage to the relativistic calculation, i.e. the use of (Ad.1) and the Dirac theory, would allow to reach the 8,173 MHz. And so, as suggested by Bethe in his article of 1947 the relativity (associated with the retardation) applied to the formula he proposed, could give the exact way of the calculation of the Lamb shifts. We have achieved a calculation of the contribution of the state P3/2 on the shift of $1/S1/2$ upon all the continuum and the result is around 10^8 MHz !. So the W_D term must be corrected by a renormalization as that was sensed by several authors only just some months after the publication of the article of Bethe. Note that even if the terms W_D, W_S, W_M could be calculated separatively with a good precision, this precision would not be right enough to give a small number as the difference of two very large numbers. The calculation level by level of $W_D + W_S - W_M$ for the high values of the continuum, which has been used until now (to our knowlegde) seems a necessity.

Part VII

Appendices

A

The Hestenes Spinor and the Pauli and Dirac Spinors

Note. In what follows, the identification of the Hamilton quaternion q and the Hestenes spinor ψ with the Pauli spinor ξ and the Dirac spinor Ψ is based on the articles [28, 44]. Perhaps, it is not the shortest one, but it allows a step-by-step conversion of Ψ into ψ and vice versa when Ψ is expressed by means of its four complex components.

The spinors ξ and Ψ are defined as columns of complex numbers, without any proper structure, and their properties follow only from the fact that the matrices σ_k and γ_μ act on these columns.

In contrast, in the real formalism, the spinors are replaced by objects already endowed with a proper structure, which do not need anything else for their employ.

The following operations of identification are complicated by the ambiguities of the complex formalism in which the "imaginary number" $\sqrt{-1}$ may correspond to two different real objects.

A.1 The Pauli Spinor as a Decomposition of the Hamilton Quaternion

Using $i = -jk$, (2.4) may be written as

$$q = u_1 - ju_2, \quad u_1 = d + kc, \quad u_2 = -b + ka. \tag{A.1}$$

Applying (2.5), we immediately obtain

$$iq = (ku_2) - j(ku_1), \quad jq = u_2 - j(-u_1), \quad kq = (ku_1), -j(-ku_2) \tag{A.2}$$

and forgetting j, let us write q as a doublet of "complex numbers" (u_1, u_2) in which the "imaginary" number $\sqrt{-1}$ is replaced by k.

Now, we consider the Pauli spinor $\xi = (u'_1, u'_2)$, where $u'_1 = d + i'c$, $u'_2 = -b + i'a$ and $i' = \sqrt{-1}$. Introducing the matrices $i'\sigma_k$, we can write

$$\mathrm{i}'\sigma_1\xi \Longleftrightarrow iq, \quad \mathrm{i}'\sigma_2\xi \Longleftrightarrow jq, \quad \mathrm{i}'\sigma_3\xi \Longleftrightarrow kq, \quad \mathrm{i}' = \sqrt{-1} \Longleftrightarrow k. \quad (A.3)$$

Note that $\mathrm{i}'u'_\alpha = u'_\alpha \mathrm{i}'$ as well as $ku_\alpha = u_\alpha k$. But this operation corresponds to the change of q into qk, i.e., the multiplication of q, *on the right* (because of the presence of j in (A.1)), by k. That cannot be seen by writing $\mathrm{i}'\xi = \xi\mathrm{i}'$, and though that the use of the standard formalism remains coherent; it forbids the interpretation of the spinor ξ otherwise than an abstract entity instead of the element of a field closely related to the group of the rotations in $E^3 = R^{3,0}$.

For convenience, we write now $i = \underline{i}e_1$, $j = \underline{i}e_2$, and $k = \underline{i}e_3$.

Multiplying in the relations (A.3) iq, jq, kq on the right by $-k = -\underline{i}e_3$ and because $-\underline{i}e_k q\underline{i}e_3 = e_k qe_3$ and $-(\mathrm{i}'\sigma_k\xi)\mathrm{i}' = \sigma_k\xi$, we obtain

$$\sigma_k\xi \Longleftrightarrow e_k qe_3. \quad (A.4)$$

A.2 The Dirac Spinor as a Decomposition of the Biquaternion

We consider a Dirac spinor $\Psi = (u'_1, u'_2, u'_3, u'_4)$. We will suppose that each u'_1, u'_2, with $\mathrm{i}' = \sqrt{-1}$, is in the form $a + \mathrm{i}'b$, and, in the purpose to be in agreement in particular with Bethe and Salpeter [5], we will suppose that each u'_3, u'_4 is in the form $\mathrm{i}''(a + \mathrm{i}'b)$ with also $\mathrm{i}'' = \sqrt{-1}$.

For the traduction in the real formalism, Ψ is a biquaternion Q that, for the while, we consider as an element of $Cl(3,0)$. We will write

$$Q = q_1 + \underline{i}q_2, \quad q_1 = u_1 - \underline{i}e_2 u_2, \quad \underline{i}q_2 = u_3 - \underline{i}e_2 u_4, \quad (A.5)$$

where each u_1, u_2 is in the form $a + \underline{i}e_3 b$, identified to $a + \mathrm{i}'b$, and each u_3, u_4 is in the form $\underline{i}(a + \underline{i}e_3 b)$, identified to $\mathrm{i}''(a + \mathrm{i}'b)$ with $\underline{i}e_3 \Leftrightarrow \mathrm{i}'$ and $\underline{i} \Leftrightarrow \mathrm{i}''$. However, the identification of the Dirac spinor Ψ to a biquaternion element ψ of $Cl^+(1,3)$ requires a justification. It lies through the consideration of the action of the Dirac matrices γ^μ upon the Dirac spinor, for the while only defined as an element of C^4, more precisely as a doublet of elements of C^2.

A.3 The Hestenes Spinor and the Dirac Matrices

As the γ^μ matrices use the σ_k ones, we are obliged to take the identification (A.4) into consideration and now to introduce e_3 inside Q.

Because $\underline{i}q = \underline{i}qe_3^2 = (q\underline{i}e_3)e_3$, one can write a quaternion $Q \in Cl(3,0)$ in the form

$$Q = q_1 + \underline{i}q_2 = q_1 + \hat{q}_2 e_3, \quad \hat{q}_2 = q_2 \underline{i}e_3, \quad (A.6)$$

with $q_1 = u_1 - \underline{i}e_2 u_2$, $\hat{q}_2 = \hat{u}_3 - \underline{i}e_2 \hat{u}_4$, and, forgetting e_3, consider Q as a doublet $(q_1, \hat{q}_2)$ of quaternions.

Note that u_3, u_4 are now replaced by $\hat{u}_3 e_3$, $\hat{u}_4 e_3$, and so u'_3, u'_4 are replaced by $\hat{u}'_3$, $\hat{u}'_4$ each one in the form $\mathrm{i}'(a+\mathrm{i}'b)$. But there is no matter in the complex formalism to distinguish $\hat{u}'_3$, $\hat{u}'_4$ from u'_3, u'_4.

Now, we consider the quaternion Q as an element ψ of $\mathrm{Cl}^+(1,3)$ and look for the concordance

$$\psi = q_1 + \hat{q}_2 e_3 \simeq (q_1, \hat{q}_2) \Longleftrightarrow \Psi = (\xi_1, \xi_2). \tag{A.7}$$

Because $e^0 \mathrm{i} q_2 e^0 = -\mathrm{i} q e^0 e_0 = -\hat{q}_2 e_3$ and $e^k = -e_k = -e_k e_0$, we obtain

$$e^0 \psi e_0 = e^0 (q_1 + \hat{q}_2 e_3) e_0 = q_1 - \hat{q}_2 e_3 \Longleftrightarrow (\xi_1, -\xi_2) = \gamma^0 \Psi,$$
$$e^k \psi e_0 = -e_k e^0 \psi e_0 = e_k \hat{q}_2 e_3 - (e_k q_1 e_3) e_3 \Longleftrightarrow (\sigma_k \xi_2, -\sigma_k \xi_1) = \gamma^\mu \Psi,$$

and so

$$e^\mu \psi e_0 \Longleftrightarrow \gamma^\mu \Psi. \tag{A.8}$$

All the concordances (3.16) are justified.

Taking ψ in the form (A.7), identifying each component to the corresponding one of Ψ, a simple calculation shows the concordance (3.17):

$$j^\mu = \bar{\Psi} \gamma^\mu \Psi \in R \Leftrightarrow e^\mu \cdot (\psi e_0 \tilde{\psi}) \in R. \tag{A.9}$$

A.4 Solution for the Central Potential Expressed by Means of the Dirac Spinors

Using (4.26), we can write

$$\phi = (g N^m_{1+\kappa} e_3 - f \underline{\mathrm{i}} N^m_{1-\kappa} e_3) \mathrm{e}^{\mathrm{i} e_3 m \varphi} \tag{A.10}$$

and the introduction of $N^m_{1-\kappa}$ is the explanation, as we are going to see, of the presence of $P^m_{l\pm 1}$, $P^{m+1}_{l\pm 1}$ in the solutions written in spinor formalism.

The expression of the biquaternion ϕ in a Dirac spinor is the following:

$$\phi = \phi_1 + \underline{\mathrm{i}} \phi_2 = u_1 - \underline{\mathrm{i}} e_2 u_2 + u_3 - \underline{\mathrm{i}} e_2 u_4. \tag{A.11}$$

Note that in (A.10), $N^m_{1+\kappa}$ and $N^m_{1-\kappa}$ are each one to be calculated with the help of (4.30), where $l = 0, 1, \ldots$ or (4.31), where $l = 1, 2, \ldots$, so with values of l which differ from 1, when in (A.11) the chosen values of l in the expression of u_1, u_2 and u_3, u_4 are the same for these two couples of numbers.

We can write, using the relations $u e_3 = -\underline{\mathrm{i}} v$, $v = e_2 \exp(\underline{\mathrm{i}} e_3 \varphi)$,

$$N e_3 = L + M u e_3 = L - M \underline{\mathrm{i}} v = L - M \underline{\mathrm{i}} e_2 \mathrm{e}^{\underline{\mathrm{i}} e_3 \varphi} \tag{A.12}$$

and so

$$g L^m_{1+\kappa} \mathrm{e}^{\mathrm{i} e_3 m \varphi} = u_1, \quad g M^m_{1+\kappa} \mathrm{e}^{\mathrm{i} e_3 (m+1) \varphi} = u_2,$$
$$-f L^m_{1+\kappa} \mathrm{e}^{\mathrm{i} e_3 m \varphi} \underline{\mathrm{i}} = u_3, \quad -f M^m_{1+\kappa} \mathrm{e}^{\mathrm{i} e_3 (m+1) \varphi} \underline{\mathrm{i}} = u_4.$$

Introducing the spherical functions $Y_l^m(\theta, \varphi)$ in which $\sqrt{-1}$ has been replaced by $\underline{i}e_3$,

$$Y_l^m(\theta, \varphi) = \frac{1}{\sqrt{2\pi}} P_l^m(\cos\theta) e^{\underline{i}e_3 m\varphi}, \qquad (A.13)$$

we can write

Case (a). $\kappa \leq -1$, $m = m' - 1/2$

For this value of κ, we apply (4.30), and for $\kappa' = -\kappa = (l+1) \geq 2$ we have to apply (4.31) in which l is to be replaced by $l+1$ (for respecting the fact that the value of l in u_1, u_2 and u_3, u_4 is the same for these two couples of numbers):

$$gL_{-l}^m = g\left[\frac{l+m+1}{2l+1}\right]^{1/2} Y_l^m = g\left[\frac{l+m'+(1/2)}{2l+1}\right]^{1/2} Y_l^{m'-1/2} = u_1,$$

$$gM_{-l}^m = g\left[\frac{l-m}{2l+1}\right]^{1/2} Y_l^m = g\left[\frac{l-m'+(1/2)}{2l+1}\right]^{1/2} Y_l^{m'+1/2} = u_2,$$

$$-fL_{l+2}^m = -f\left[\frac{l+1-m}{2l+3}\right]^{1/2} Y_{l+1}^m \underline{i} = -f\left[\frac{l-m'+(3/2)}{2l+3}\right]^{1/2} Y_{l+1}^{m'-1/2}\underline{i} = u_3,$$

$$fM_{l+1}^m = f\left[\frac{l+m+2}{2l+3}\right]^{1/2} Y_{l+1}^{m+1}\underline{i} = f\left[\frac{l+m'+(3/2)}{2l+3}\right]^{1/2} Y_{l+1}^{m'+1/2}\underline{i} = u_4.$$

Case (b). $\kappa \geq 1$, $m = m' - 1/2$

For this value of κ one applies (4.31), and for $\kappa' = -\kappa = -l \leq 0$ we have to apply (4.30) in which l is to be replaced by $l-1$ (for respecting the fact that the value of l in u_1, u_2 and u_3, u_4 is the same for these two couples of numbers):

$$gL_{l+1}^m = g\left[\frac{l-m}{2l+1}\right]^{1/2} Y_l^m = g\left[\frac{l-m'+(1/2)}{2l+1}\right]^{1/2} Y_l^{m'-1/2} = u_1,$$

$$gM_{l+1}^m = -g\left[\frac{l+m+1}{2l+1}\right]^{1/2} Y_{l+1}^m = -g\left[\frac{l+m'+(1/2)}{2l+1}\right]^{1/2} Y_l^{m'+1/2} = u_2,$$

$$-fL_{1-l}^m = -f\left[\frac{l+m}{2l-1}\right]^{1/2} Y_{l-1}^m\underline{i} = -f\left[\frac{l+m'-(1/2)}{2l-1}\right]^{1/2} Y_{l-1}^{m'-1/2}\underline{i} = u_3,$$

$$-fM_{1-l}^{m+1} = -f\left[\frac{l-m-1}{2l+3}\right]^{1/2} Y_{l-1}^{m+1}\underline{i} = -f\left[\frac{l-m'-(1/2)}{2l-1}\right]^{1/2} Y_{l-1}^{m'+1/2}\underline{i} = u_4.$$

So, one can obtain the spinor expression of the solutions (14.3) and (14.4) for the case (a), and (14.7) for the case (b) of Bethe and Salpeter [5] (with a change of sign for u_3 and u_4 due to a change of the convention of the orientation of the plane (e_1, e_2)).

B

The Real Formalism and the Invariant Entities

B.1 Properties of the Hestenes Spinor

Let $\psi \in \mathrm{Cl}^+(M)$ such that $\psi\tilde{\psi} \neq 0$. We can write

$$\psi\tilde{\psi} \in \mathrm{Cl}^+(M) \Rightarrow \psi\tilde{\psi} = \lambda + B + \underline{\mathrm{i}}\mu, \quad \lambda, \mu \in R, B \in \wedge^2 M$$

and from $(\psi\tilde{\psi})\tilde{\ } = \psi\tilde{\psi}$, $\tilde{B} = -B$, $\tilde{\underline{\mathrm{i}}} = \underline{\mathrm{i}}$, we deduce $B = 0$ and

$$\psi\tilde{\psi} = \lambda + \underline{\mathrm{i}}\mu = \rho e^{\underline{\mathrm{i}}\beta}, \quad \rho > 0, \ \beta \in R,$$

then [35]

$$\frac{\psi\tilde{\psi}}{\rho e^{\underline{\mathrm{i}}\beta}} = 1, \quad R = \frac{\psi}{\sqrt{\rho}e^{\underline{\mathrm{i}}\beta/2}} \Rightarrow \quad \psi = \sqrt{\rho}e^{\underline{\mathrm{i}}\beta/2}R, \quad R\tilde{R} = \tilde{R}R = 1. \tag{B.1}$$

So, R verifies $\tilde{R} = R^{-1}$ and corresponds to a representation of $SO^+(1,3)$ in $\mathrm{Cl}^+(M)$. Because $a\underline{\mathrm{i}} = -\underline{\mathrm{i}}a$ if $a \in M$,

$$a \in M \Rightarrow b = Ra\tilde{R} \in M, \quad \psi a\tilde{\psi} = \rho b \in M. \tag{B.2}$$

If ψ is associated with a Galilean frame $\{e_\mu\}$, all the properties of invariance met in the Dirac theory are immediately deduced, in particular the one of the Dirac current

$$j = \psi e_0 \tilde{\psi} = \rho v, \quad \rho > 0, \quad v^2 = 1, \tag{B.3}$$

where ρ expresses the invariant probability density.

B.2 The Proper Angular Momentum or Bivector Spin

The bivector spin σ is deduced from (B.2) by (see [36], (1.6))

$$\sigma = \frac{\hbar c}{2}(n_2 \wedge n_1), \quad n_k = Re_k\tilde{R} = \frac{\psi e_k \tilde{\psi}}{\rho}. \tag{B.4}$$

The change $e_2 e_1$ into $e_1 e_2$ in the Dirac equation inverses the orientation of the bivector spin.

B.3 The Energy–Momentum Vector

The energy–momentum vector is the value $p = T(v) \in M$ for v of the energy–momentum (Tetrode) tensor ρT. We have shown (see [9]) that it is in the form

$$p = \frac{\hbar c}{2} \varpi - eA, \quad \varpi_\mu = (\partial_\mu n_2) \cdot n_1 = -n_2 \cdot (\partial_\mu n_1), \tag{B.5}$$

where ϖ may be interpreted as expressing the infinitesimal rotation upon itself of the "spin plane" $\pi(x) = (n_1, n_2)$. Also, we have verified in [9] with this geometrical interpretation of ϖ that the energy

$$E = \frac{\hbar c}{2} \varpi \cdot e_0 = \frac{\hbar c}{2} \varpi^0 = p^0 + eA^0 \tag{B.6}$$

of the electron in the Galilean frame $\{e_\mu\}$ where the nucleus is at rest is effectively the E of (3.5) for the hydrogenic atoms.

So, the total angular momentum

$$J = x \wedge p + \sigma \tag{B.7}$$

implies not only, inside σ, the direction of the plane $\pi(x)$, but also, into p, the proper infinitesimal rotation of this plane.

We recall that the group of the finite rotations upon itself of this plane corresponds to the gauge $U(1)$ in the theory of the electron, as it simply deduced from (B.2) (see [35] but also, in the complex formalism, [31, 39]).

So, three fundamental properties in the theory of the electron, the energy, the spin, and the gauge, are directly related to a plane orthogonal to the Dirac current.

Note. It is to emphasize that the above entities are independent of the probability density ρ, and so are relevant of the part D_I (four real scalar equations) of the Dirac equation D which is independent of ρ. Let D_{II} (four real scalar equations) be the part of D which depends on ρ. About the role of the density ρ with respect to these entities, we have established in [11] the following theorem.

D_{II} is implied by D_I and the three conservation relations

$$\partial_\mu(\rho v^\mu) = 0, \quad \partial_\mu(\rho T^{\mu\nu}) = \rho f^\nu, \quad \partial_\mu(\rho S^{\mu\nu\xi}) = \rho(T^{\xi\nu} - T^{\nu\xi}),$$

where $f \in M$ is the Lorentz force, $S = v \wedge \sigma \in \wedge^3 M$.

C

The Total Angular Momentum Operator

In the usual presentation of the Dirac theory, one considers the following operator, here expressed in a STA form

$$L = \sigma_0 \cdot (x \wedge \partial) - \frac{1}{2}\sigma_0 \quad (\sigma_0 = e_1 \wedge e_2, \ \partial = e^\mu \partial_\mu). \tag{C.1}$$

Writing $x \wedge \partial = (x^0 e_0 + rn) \wedge (e^0 \partial_0 + e^k \partial_k)$, because $\sigma_0 \cdot (n \wedge e^0) = 0$, and noting that $e^k = -e_k$, $a, b \in M$, $a.e_0 = 0 = b \cdot e_0 \Rightarrow a \wedge b = -\boldsymbol{a} \wedge \boldsymbol{b}$, $e_1 \wedge e_2 = -\underline{i}e_3$, we obtain

$$L = -\underline{i}e_3 \cdot (\boldsymbol{r} \wedge \nabla) + \frac{1}{2}\underline{i}e_3 = \partial_\varphi + \frac{1}{2}\underline{i}e_3, \quad \underline{i}e_3 = e_2 \wedge e_1. \tag{C.2}$$

Taking S as in (4.15), because $\partial_\varphi \boldsymbol{u}e_3 = \boldsymbol{v}e_3 = \boldsymbol{u}e_3\underline{i}e_3$, we obtain

$$LS = \left(m + \frac{1}{2}\right)S\underline{i}e_3, \quad \underline{i}e_3 = e_2 \wedge e_1. \tag{C.3}$$

Applying (A1.10), we have

$$L\psi = \left(m + \frac{1}{2}\right)\psi e_2 e_1, \quad e_2 e_1 = e_2 \wedge e_1, \tag{C.4}$$

from which we deduce the relation

$$\hbar c(L\psi)\psi^{-1} = \left(m + \frac{1}{2}\right)\hbar c(n_2 \wedge n_1), \quad m \in Z, \tag{C.5}$$

which implies the bivector spin $\sigma = (\hbar c/2)n_2 \wedge n_1$ and also the magnetic number $m \in Z$.

D

The Main Properties
of the Real Clifford Algebras

The field $H = \text{Cl}^+(3,0)$ of the Hamilton quaternions and the ring $\text{Cl}(3,0)$ of the Clifford biquaternions are relevant of the general theory of the Clifford algebra $\text{Cl}(E) = \text{Cl}(p, n-p)$ associated with an euclidean space $E = R^{p,n-p}$. They correspond to the initial construction of the Clifford algebras. Especially, the field of the Hamilton quaternions plays an important role in the solution of the central potential problem.

The general definition and properties of the Clifford algebras may be seen in [34]. We simply mention here that $\text{Cl}(E)$ is an associative real algebra acting upon the elements of R and the vectors of E, in relation with the Grassmann algebra $\wedge E$.

We recall that the Grassmann (or exterior) algebra $\wedge R^n$ of R^n is an associative algebra generated by R and the vectors of R^n such that the Grassmann product $a_1 \wedge a_2 \wedge \cdots \wedge a_p$ of vectors $a_k \in R^n$ is null if and only if the a_k are linearly dependent. If this product is non-null, it is called a simple (or decomposable) p-vector and owns the geometrical meaning of a p-paralleloid (a parallelogram if $p = 2$). The linear combination of simple p-vectors is called a p-vector, and the set of the p-vectors is a sub-space, denoted $\wedge^p R^n$, of $\wedge R^n$.

One deduces easily that $\wedge R^n$ is the direct product of the sub-spaces $\wedge^p R^n$ ($p = 0, 1, \ldots, n$), each one of dimension C_n^p, with $\wedge^0 R^n = R$, and so $\dim(\wedge R^n) = 2^n$.

Certainly, $\wedge R^n$ is the first algebra to be associated with R^n because it is based on the notion of linear independence of vectors, which is in the foundation of the definition of the vector spaces.

The elements of $\wedge^p R^n$ are presented in Physics as "tensors completely antisymmetric of rank p," but their use needs in this case the resort to a frame of R^n, which is not necessary. Associated with a signature of R^n, they have generally a physical meaning, as for example the electromagnetic field $F \in \wedge^2 M$.

The interest of the use of a real Clifford algebra $\text{Cl}(E)$ of an euclidean space E lies in the fact that the elements of this algebra are identified to the ones of $\wedge E$. Then, $\text{Cl}(E)$ not only contains the geometrical elements of the space E but also can express the transformations of these elements, and so by means of objects which are also geometrical elements of E.

Let us denote $a \cdot b$ the scalar product of two vectors of a space $E = R^{p,n-p}$.

The Clifford product of two elements A, B of $\mathrm{Cl}(E)$ is denoted AB and verifies the fundamental relation

$$a^2 = a \cdot a, \quad \forall a \in E, \tag{D.1}$$

from which we deduce

$$(a + b)^2 = a^2 + ab + ba + b^2 = (a + b) \cdot (a + b) = a \cdot a + 2a \cdot b + b \cdot b$$

and so

$$a \cdot b = \frac{1}{2}(ab + ba). \tag{D.2}$$

Now,

$$ab = \frac{1}{2}(ab + ba) + \frac{1}{2}(ab - ba)$$

and identifying $(ab - ba)/2$ to $a \wedge b$, a convention that nothing forbids, one can write

$$ab = a \cdot b + a \wedge b \quad (a, b \in E), \tag{D.3}$$

in such a way that

$$a \cdot b = 0 \implies ab = a \wedge b = -b \wedge a = -ba. \tag{D.4}$$

We will not detail here the identification of the elements of $\mathrm{Cl}(E)$ to elements of $\wedge E$ (see [34]). We only mention a property we need: if p vectors $a_i \in E$ are orthogonal, their Clifford product verifies

$$a_1 \ldots a_p = a_1 \wedge \cdots \wedge a_p \quad (a_k \in E, \ a_i \cdot a_j = 0, \ \text{if } i \neq j). \tag{D.5}$$

The even sub-algebra $\mathrm{Cl}^+(E)$ of $\mathrm{Cl}(E)$ is composed by the sums of scalars and elements $a_1 \ldots a_p$ such that $p = 2q$.

One can easily deduce from (D.5) that, using an orthonormal frame of E, the corresponding frame of $\mathrm{Cl}(E)$ may be identified to the frame of $\wedge E$ and that $\dim(\mathrm{Cl}(E)) = \dim(\wedge E) = 2^n$, and $\dim(\mathrm{Cl}^+(E)) = 2^{n-1}$.

So the use of $\mathrm{Cl}(3, 0)$ or $\mathrm{Cl}^+(1, 3)$ allows one to replace the manipulation of the Pauli and Dirac matrices *and spinors* by vectors of $E^3 = R^{3,0}$ or $M = R^{1,3}$ with a simple rule of an associative product on these vectors without the obligation of resorting to a frame of E^3 or M, which is a necessity in the complex spinors formalism.

Note. Equation (D.3) is accepted with difficulty by many physicists. What does that mean the sum of a scalar and a bivector?

One may identify $\text{Cl}^+(E^2)$ with the so well-known (and often used needlessly) field C of the complex numbers by the following relations:

$$e_2 \wedge e_1 = e_2 e_1, \quad (e_2 e_1)^2 = -1, \quad e_2 e_1 \Longleftrightarrow i \qquad (\text{D.6})$$
$$\Rightarrow ab = a \cdot b + a \wedge b = \rho(\cos\theta + e_2 \wedge e_1 \sin\theta) \Rightarrow e^{e_2 e_1 \theta} \Longleftrightarrow e^{i\theta},$$

which show that ab may be associated with a rotation in the plane E^2.

In E^3, ab is a Hamilton quaternion in which, associated with a frame of E^3, three *different* bivectors appear whose square is equal to -1, and may be associated, as it is well known, with a rotation in E^3.

So, the use of $\text{Cl}^+(E^2)$ and $\text{Cl}^+(E^3)$ may replace the use of $U(1)$ and $SU(2)$.

What does that mean? That means that the imaginary number $\sqrt{-1}$ is a symbol which hides a geometrical object. And, this object may be different following the use of this symbol: different bivectors, but furthermore, objects of different geometrical nature as shown in (2.15).

In a general way, in all $\text{Cl}(E)$, the Clifford product $a_1 a_2 \ldots a_p$, where $(a_k \in E, a_k^2 \neq 0)$, may be associated with an isometry in the space E, and so $\text{Cl}(E)$ may replace the general theory of the representations of the orthogonal group $O(E)$ in complex spaces.

Consider the relation

$$y = -bxb, \quad b^2 = b \cdot b \neq 0, \quad b, x \in E. \qquad (\text{D.7})$$

Let the decomposition $x = x^\perp + x^\parallel$, where $x^\parallel$ and $x^\perp$ are parallel and orthogonal to b, respectively. Equation (D.4) allows one to write

$$y = b^2(x^\perp - x^\parallel) \in E,$$

and we see that the transformation $x \in E \rightarrow y \in E$ is a symmetry with respect to the hyperplan orthogonal to b, followed by the multiplication by the scalar b^2. The relation $z = -aya = abxba$ is a rotation followed by the multiplication by the scalar $a^2 b^2$. So, the relation

$$y = (-1)^p I_p x I_p^{-1}, \; I_p = a_1 a_2 \ldots a_p, I_p^{-1} = a_p^{-1} \ldots a_2^{-1} a_1^{-1}, a_k^2 \neq 0, a_k^{-1} = \frac{a_k}{a_k^2}$$
$$(\text{D.8})$$

defines an isometry in E. We see the tight links between the orthogonal group $O(E)$ of the space E and its Clifford algebra $\text{Cl}(E)$. There exists a proof using $\text{Cl}(E)$ (H. Krüger 1998, private communication), quite different of the one of Cartan-Dieudonné, of the Elie Cartan theorem by which all isometry in E is the product of symmetries each one with respect to a non-isotropic hyperplan.

A combination of $a_1 a_2 \ldots a_p$ does not give necessarily an isometry but some of them can lead to euclidean transformations (obtained with difficulty by means of the complex formalism) whose role in Quantum Mechanics is important (as the one associated with the angle of Yvon–Takabayasi–Hestenes β of (AI.2.1)) (see [35, 56, 57]).

E

The Expression of the Transition Current

We denote

$$\psi_k = T_k e^{\underline{i} e_3 (m_k \varphi - (E_k/\hbar c) x^0)}.$$

Since $(\underline{i}e_3)\tilde{} = -\underline{i}e_3$ and $e_0 \underline{i} e_3 = \underline{i} e_3 e_0$, using the notations of Chap. 4 we can write

$$\psi_1 e_0 \tilde{\psi}_2 = \cos(\epsilon\varphi + \omega x^0) T_1 e_0 \tilde{T}_2 + \sin(\epsilon\varphi + \omega x^0) T_1 \underline{i} e_3 e_0 \tilde{T}_2,$$

$$\psi_2 e_0 \tilde{\psi}_1 = \cos(\epsilon\varphi + \omega x^0) T_2 e_0 \tilde{T}_1 - \sin(\epsilon\varphi + \omega x^0) T_2 \underline{i} e_3 e_0 \tilde{T}_1.$$

We use the relation

$$X \in \wedge^1 M + \wedge^3 M \implies X + \tilde{X} = [X]_v,$$

where $[Y]_v$ means the vector part of $Y \in \mathrm{Cl}(M)$. So, (5.8) is proved with the vectors j_I and j_{II} in the form

$$j_I = T_1 e_0 \tilde{T}_2 + T_2 e_0 \tilde{T}_1 = 2[T_1 e_0 \tilde{T}_2]_v$$

and because $(\underline{i}e_3 e_0)\tilde{} = -\underline{i}e_3 e_0$

$$j_{II} = T_1 \underline{i} e_3 e_0 \tilde{T}_2 - T_2 \underline{i} e_3 e_0 \tilde{T}_1,$$

$$j_{II} = T_1 \underline{i} e_3 e_0 \tilde{T}_2 + [T_1 \underline{i} e_3 e_0 \tilde{T}_2]\tilde{} = 2[T_1 \underline{i} e_3 e_0 \tilde{T}_2]_v.$$

(a) The spatial parts $\mathbf{j}_I$ and $\mathbf{j}_{II}$ of j_I and j_{II} may be calculated in the following way.

(1) Since $e_0 \underline{i} = -\underline{i} e_0$, we can write

$$j_I = 2[(g_1 \mathbf{N}_1 \mathbf{e}_3 - f_1 \underline{i} \mathbf{n} \mathbf{N}_1)(g_2 \mathbf{e}_3 \mathbf{N}_2 + f_2 \mathbf{N}_2 \mathbf{n} \underline{i}) e_0]_v$$

and since $e_0^2 = 1$, we obtain

$$\mathbf{j}_I = [j_I e_0]_V = 2[(g_1 \mathbf{N}_1 \mathbf{e}_3 - f_1 \mathbf{i} \mathbf{n} \mathbf{N}_1)(g_2 \mathbf{e}_3 \mathbf{N}_2 + f_2 \mathbf{N}_2 \mathbf{n}) \mathbf{i}]_V,$$

where $[X]_V$ means the vector part of $X \in \mathrm{Cl}(E^3)$. Equation (5.11) may be deduced without difficulty.

The coefficients of $g_1 g_2$ and $f_1 f_2$ are null as parts of elements of $\mathrm{Cl}^+(E^3)$, and so sums of a scalar and a bivector.

We give, for example, the calculation of the coefficient of $f_1 g_2$.

Using

$$\mathbf{ab} = \mathbf{a} \cdot \mathbf{b} + \mathbf{a} \wedge \mathbf{b}, \quad \mathbf{a} \wedge \mathbf{b} = \underline{\mathbf{i}}\,(\mathbf{a} \times \mathbf{b}),$$

we may write

$$-2[\underline{\mathbf{i}} \mathbf{n} \mathbf{N}_1 \mathbf{e}_3 \mathbf{N}_2]_V = 2[-(\underline{\mathbf{i}}(\mathbf{n} \cdot \mathbf{N}_1) + \mathbf{n} \times \mathbf{N}_1)(\mathbf{e}_3 \cdot \mathbf{N}_2 + \underline{\mathbf{i}}(\mathbf{e}_3 \times \mathbf{N}_2)]_V$$

with

$$[(\mathbf{n} \times \mathbf{N}_1)\underline{\mathbf{i}}(\mathbf{e}_3 \times \mathbf{N}_2)]_V = -\underline{\mathbf{i}}((\mathbf{n} \times \mathbf{N}_1) \wedge (\mathbf{e}_3 \times \mathbf{N}_2)) = 0,$$

because $\mathbf{n} \times \mathbf{N}_1$ and $\mathbf{e}_3 \times \mathbf{N}_2$ are each one colinear to $\mathbf{v}$, and so we obtain

$$-2[\underline{\mathbf{i}} \mathbf{n} \mathbf{N}_1 \mathbf{e}_3 \mathbf{N}_2]_V = 2[(\mathbf{n} \cdot \mathbf{N}_1)(\mathbf{e}_3 \times \mathbf{N}_2) + (\mathbf{e}_3 \cdot \mathbf{N}_2)(\mathbf{n} \times \mathbf{N}_1)]$$

$$= 2[(L_1 \cos\,\theta + M_1 \sin\,\theta)M_2 + L_2(M_1 \cos\,\theta - L_1 \sin\,\theta)]\mathbf{v}.$$

The coefficient of $f_2 g_1$ is obtained in the same way and the expression of $\mathbf{j}_I$ is proved.

(2) We can write

$$\mathbf{j}_{II} = [j_{II} e_0]_V = 2[(g_1 \mathbf{N}_1 \mathbf{e}_3 - f_1 \underline{\mathbf{i}} \mathbf{n} \mathbf{N}_1)\underline{\mathbf{i}} \mathbf{e}_3 (g_2 \mathbf{e}_3 \mathbf{N}_2 + f_2 \mathbf{N}_2 \mathbf{n} \underline{\mathbf{i}})]_V.$$

The coefficients of $g_1 g_2$ and $f_1 f_2$ are null as vector parts of the sum of elements of $\wedge^0 E^3$, $\wedge^2 E^3$ and $\wedge^3 E^3$. Then

$$\mathbf{j}_{II} = 2[f_1 g_2 \mathbf{n} \mathbf{N}_1 \mathbf{N}_2 - g_1 f_2 \mathbf{N}_1 \mathbf{N}_2 \mathbf{n}]_V.$$

The coefficient of $f_1 g_2$ is, because $\mathbf{n}(\mathbf{u} \wedge \mathbf{e}_3) = \mathbf{n}(\mathbf{u} \mathbf{e}_3) = -\mathbf{w}$,

$$\mathbf{n} \mathbf{N}_1 \mathbf{N}_2 = \mathbf{n}(\mathbf{N}_1 \cdot \mathbf{N}_2 + \mathbf{N}_1 \wedge \mathbf{N}_2) = (L_1 L_2 + M_1 M_2)\mathbf{n} + (L_1 M_2 - L_2 M_1)\mathbf{w}.$$

The coefficient of $f_2 g_1$ is calculated in the same way and (5.12) is proved.

(b) Concerning J_I^0 and J_{II}^0, we obtain

$$J_I^0 = 2(g_1 g_2 + f_1 f_2)\mathbf{N}_1 \cdot \mathbf{N}_2, \quad J_{II}^0 = 0$$

and since $\mathbf{N}_1 \cdot \mathbf{N}_2$ contains terms in the form $P_j^\mu P_k^\mu$ with $j \neq k$, and so $\int_0^\pi P_j^\mu P_k^\mu \sin\theta d\theta = 0$, (5.4) is proved.

F

Conservation of the Charge Transition Current

Let $a \in M$, $X = V + T$, where $V \in \wedge^1 M = M$, $T \in \wedge^3 M$. We can write

$$a \cdot V = [aX]_S, \quad a \cdot V = V \cdot a = [Xa]_S, \quad [Y]_S + [Z]_S = [Y + Z]_S,$$

where $[Y]_S$ means the scalar part of $Y \in \mathrm{Cl}(M)$.

So if $L_{kj} = \psi_k e_0 \tilde{\psi}_j$, we can write

$$e^\mu \cdot \partial_\mu L_{12} = [e^\mu \partial_\mu \psi_1 e_0 \tilde{\psi}_2]_S + [\psi_1 e_0 \partial_\mu \tilde{\psi}_2 e^\mu]_S$$

or taking into account (3.19)

$$e^\mu \cdot \partial_\mu L_{12} = \frac{1}{\hbar c}[(mc^2 \psi_1 e_0 + qA\psi_1)e_1 e_2 e_0 \tilde{\psi}_2 + \psi_1 e_0 e_2 e_1(mc^2 e_0 \tilde{\psi}_2 + q\tilde{\psi}_2 A)]_S = 0,$$

since $e_1 e_2 = -e_2 e_1$ and since the term containing q is in the form $[AX]_S - [XA]_S = 0$.

In the same way, one can write $e^\mu \cdot \partial_\mu L_{21} = 0$ and so $e^\mu \cdot \partial_\mu j_{12} = 0$.

G

An Approximation Method
for Time-Dependent Perturbation

For justifying the form (8.1)–(8.4) of the matrix elements, we will follow the method of perturbation described in the Sects. 29 and 32 of [50]. But here, this method will be directly applied to the Dirac theory of the electron and with the use of the real formalism.

Let us consider a wave function ψ in the form

$$\psi(x^0, \mathbf{r}) = \sum_n a_n(x^0)\psi_n(x^0, \mathbf{r}), \quad a_n(x^0) \in R, \tag{G.1}$$

where each ψ_n is the solution (4.5) of (3.19) for an electron in an hydrogenic atom in a state of energy E_n.

We suppose that, at a time $t = x_0/c$, a potential $A = A^k e_k$ is added to the central potential $A^0 e_0$ such that, as in (4.6), $eA^0 = V(r)$, $(e = -q > 0)$.

Then, the function ψ obeys the relation

$$e^\mu \partial_\mu \psi = -\frac{1}{\hbar c}(mc^2\psi e_0 - e(A^0 e_0 + A^k e_k \psi)\underline{i}e_3). \tag{G.2}$$

We suppose furthermore that this change of potential will affect only the coefficients a_n and not the functions ψ_n. Such a supposition may be justified by the fact that the effect of the perturbative potential is the passage from a state of energy E_j to a state of energy E_k.

We can write

$$e^\mu \partial_\mu \psi = \sum_n (\dot{a}_n e^0 \psi_n + a_n e^\mu \partial_\mu \psi_n), \tag{G.3}$$

where $\dot{a}_n(x^0)$ means the derivative of $a_n(x^0)$ with respect to x^0.

Since ψ_n is a solution in the absence of the perturbating potential A, we deduce from (G.2) and (G.3), multiplying on the left by $e^0 = e_0$,

$$\sum_n \dot{a}_n \psi_n = \frac{e}{\hbar c} \sum_n A^k e_0 e_k a_n \psi_n \underline{i}e_3. \tag{G.4}$$

Considering the transition current between a state m and a state n

$$j_{mn} = \psi_m e_0 \tilde{\psi}_n + \psi_n e_0 \tilde{\psi}_m \in M \quad \text{and} \quad j_{mn}^0 = j_{mn} \cdot e_0, \qquad (\text{G.5})$$

we deduce easily

$$\sum_n \dot{a}_n j_{mn}^0 = \frac{e}{\hbar c} A^k \sum_n a_n [e_0 e_k \psi_n \underline{\mathrm{i}} e_3 \tilde{\psi}_m + \psi_m e_3 \underline{\mathrm{i}} \tilde{\psi}_n e_k e_0] \cdot e_0. \qquad (\text{G.6})$$

Using the relation

$$[e_0 e_k X + \tilde{X} e_k e_0] \cdot e_0 = [e_0^2 e_k X]_S + [\tilde{X} e_k e_0^2]_S = e_k \cdot [X + \tilde{X}],$$

we deduce

$$\sum_n \dot{a}_n j_{mn}^0 = \frac{e}{\hbar c} A^k e_k \cdot \sum_n a_n (\psi_n \underline{\mathrm{i}} e_3 \tilde{\psi}_m + \psi_m e_3 \underline{\mathrm{i}} \tilde{\psi}_n) \qquad (\text{G.7})$$

and using the same methods as in (G.1)

$$\sum_n \dot{a}_n j_{mn}^0 = \frac{e}{\hbar c} \mathbf{A} \cdot \sum_n a_n \mathbf{g}_{mn}, \quad \mathbf{A} = A^k \mathbf{e}_k \qquad (\text{G.8})$$

with

$$\mathbf{g}_{mn} = \sin(\omega_{mn} x^0) \mathbf{j}_{1,mn} - \cos(\omega_{mn} x^0) \mathbf{j}_{2,mn}, \quad \omega_{mn} = \frac{E_n - E_m}{\hbar c}, \qquad (\text{G.9})$$

where $\mathbf{j}_{1,mn}$ and $\mathbf{j}_{2,mn}$ are defined as in (5.10) and (5.11). So, a new time-periodic vector $\mathbf{g}$ appears related to the spatial component $\mathbf{j}$ ((5.9) of the transition current between two states).

Now, we use (4.18) and (5.4)

$$\int j_{mn}^0 \mathrm{d}\tau = \delta_{mn} \qquad (\text{G.10})$$

and we obtain

$$\dot{a}_m(x^0) = \frac{e}{\hbar c} \int \mathbf{A} \cdot \sum_n a_n(x^0) \mathbf{g}_{mn}(x^0, \mathbf{r}) \mathrm{d}\tau. \qquad (\text{G.11})$$

We are going to use the perturbation approximation method (see Sect. 29 of [50]) which consists in replacing $\mathbf{A}$ by $\lambda \mathbf{A}$ and expressing each a_n as power series in λ:

$$a_n = a_n^{(0)} + \lambda a_n^{(1)} + \lambda^2 a_n^{(2)} + \cdots . \qquad (\text{G.12})$$

Each term of the series corresponds to an order of approximation. We will consider only the first order.

Equating the coefficients of equal power of λ, we obtain

$$\dot{a}_m^{(1)}(x^0) = \frac{e}{\hbar c} \int \mathbf{A} \cdot \sum_n c_n \mathbf{g}_{mn}(x^0, \mathbf{r}) \mathrm{d}\tau, \qquad (G.13)$$

where each c_n is a constant. Indeed, (G.11) and (G.12) (with $\mathbf{A}$ replaced by $\lambda \mathbf{A}$) give

$$\dot{a}_m^{(0)}(x^0) = 0 \ \Rightarrow a_m^{(0)} = c_m, \qquad (G.14)$$

and as (G.13) may be written for all index n, we can write $a_n^{(0)} = c_n$ for all n.

Now, we consider two particular states j and k. The first one will be considered as the state of the electron before the beginning of the perturbation and the second one as the expected final state.

Choosing the constants c_n such that $c_n = \delta_{jk}$, we obtain

$$\dot{a}_j^{(1)}(x^0) = \frac{e}{\hbar c} \int \mathbf{A} \cdot \mathbf{g}_{jk}(x^0, \mathbf{r}) \mathrm{d}\tau. \qquad (G.15)$$

H

Perturbation by a Plane Wave

In this case, the potential $\mathbf{A}$ is such that

$$e\mathbf{A} = U\cos(\mathbf{k}\cdot\mathbf{r} - \varpi x^0 + \xi)\mathbf{L}, \tag{H.1}$$

$$\mathbf{k} = \varpi\mathbf{K}, \quad \mathbf{K}^2 = \mathbf{L}^2 = 1, \quad \mathbf{K}\cdot\mathbf{L} = 0,$$

where U is a constant and ξ is a phase constant.

The way that we follow here differs partially from the one of Schiff [50] but leads to the same conclusion. It is applied here directly to the Dirac theory of the electron instead of the Schrödinger one.

We denote now $j = 1, k = 2$ and use the notations of Sects. 4.2 and 7.1 with $\omega = (E_2 - E_1)/\hbar c$.

A simple calculation shows that (G.15) becomes

$$\dot{a}_1^{(1)}(x^0) = \alpha U \mathbf{L} \cdot \left[\int \cos(\mathbf{k}\cdot\mathbf{r} - \varpi x^0 + \xi)(\sin(\omega x^0)\mathbf{j}_1 - \cos(\omega x^0)\mathbf{j}_2)\mathrm{d}\tau\right]. \tag{H.2}$$

Taking into account (8.3) (in such a way that the terms containing $\sin(\mathbf{k}\cdot\mathbf{r})$ may be omitted in the calculation), we obtain without difficulty, denoting $\Omega = \omega - \varpi$,

$$\dot{a}_1^{(1)}(x^0) = 2\alpha U \mathbf{L} \cdot [\sin(\Omega x^0 + \xi)\mathbf{T}_1^{\perp}(\mathbf{k}) - \cos(\Omega x^0 + \xi)\mathbf{T}_2^{\perp}(\mathbf{k})] + I, \tag{H.3}$$

where I implies terms containing $\omega + \varpi$ which will not be taken into account (see Sect. 35 of [50]) because the probability of finding the system in the state 2 after the perturbation requires that $\omega - \varpi$ is close to zero.

We consider the integration with respect to x^0:

$$a_1^{(1)}(x^0) = \int_0^{x^0} \dot{a}_1^{(1)}(x)\mathrm{d}x, \tag{H.4}$$

which gives

$$a_1^{(1)}(x^0) = \frac{2\alpha U}{\Omega} \mathbf{L} \cdot [(\cos \xi - \cos(\xi + \Omega x^0))\mathbf{T}_1^{\perp}(\mathbf{k}) + (\sin \xi - \sin(\xi + \Omega x^0))\mathbf{T}_2^{\perp}(\mathbf{k})].$$

(H.5)

The average of $[a_1^{(1)}(x^0)]^2$ upon the phase factor ξ

$$\langle [a_1^{(1)}(x^0)]^2 \rangle = \frac{1}{2\pi} \int_0^{2\pi} [a_1^{(1)}(x^0)]^2 d\xi$$

leads to the formula

$$\langle [a_1^{(1)}(x^0)]^2 \rangle = 8\alpha^2 U^2 \left([\mathbf{L} \cdot \mathbf{T}_1^{\perp}(\mathbf{k})]^2 + [\mathbf{L} \cdot \mathbf{T}_2^{\perp}(\mathbf{k})]^2 \right) \frac{\sin^2((\omega - \varpi)x^0)/2)}{(\omega - \varpi)^2}.$$

(H.6)

The average upon all the directions of the vector $\mathbf{L}$ gives (see (7.5))

$$\langle\langle [a_1^{(1)}(x^0)]^2 \rangle\rangle = 4\alpha^2 U^2 ([\mathbf{T}_1^{\perp}(\mathbf{k})]^2 + [\mathbf{T}_2^{\perp}(\mathbf{k})]^2) \frac{\sin^2((\omega - \varpi)x^0)/2)}{(\omega - \varpi)^2}. \quad (H.7)$$

These formulas are similar to the one of (35.16) of [50]. They show that the probability for the transition from the state of energy E_1 to the state of energy E_2 is maximum (see Fig. 27 of [50]) when $\varpi = \omega$. Then, $\mathbf{k} = \omega\mathbf{K}$ and the numbers $\mathbf{L} \cdot \mathbf{T}_j^{\perp}(\mathbf{k})$ correspond to the matrix elements as they are considered in Chap. 8.

References

1. A.O. Barut, Y. Salamin, Phys. Rev. A **37**, 2284 (1988)
2. A.O. Barut, J.F. Van Huele, Phys. Rev. A **32**, 3187 (1985)
3. K. Bechert, Phys. Rev. **6**, 700 (1930)
4. H. Bethe, Phys. Rev. **72**, 339 (1947)
5. H. Bethe, E. Salpeter, *Quantum Mechanichs for One- and Two-Electron Atoms* (Springer, Berlin, 1957)
6. J.B. Bjorken, S.D. Drell, *Relativistic Quantum Fields* (Mc Graw-Hill, New York, 1964)
7. B. Blaive, R. Boudet, Ann. Fond. L. de Broglie bf 14, 147 (1989)
8. B. Blaive, A.O. Barut, R. Boudet, J. Phys. B **24**, 3121 (1991)
9. R. Boudet, C. R. Acad. Sci. *(Paris)* A **278**, 1063 (1974)
10. R. Boudet, C. R. Acad. Sci. *(Paris)* A **280**, 1365 (1975)
11. R. Boudet, J. Math. Phys. **28**, 718 (1985)
12. R. Boudet, *New Frontiers in Quantum Electrodynamics and Quantum*, ed. by A.O. Barut (Plenum, New York, 1990), p. 443
13. R. Boudet, Found. Phys. **23**, 1367 (1993)
14. R. Boudet, Banach Center Publ. **37**, 337 (1996)
15. R. Boudet, Found. Phys. **29**, 29 (1999)
16. R. Boudet, Found. Phys. **29**, 49 (1999)
17. R. Boudet, B. Blaive, Ann. Fond. L. de Broglie **23**, 83 (1998)
18. R. Boudet, B. Blaive, Found. Phys. **30**, 1283 (2000)
19. R. Boudet, B. Blaive, S. Geniyes, M. Vanel, Int. J. Theor. Phys. **8**, 203 (2002)
20. E. Chupp, L. Dotchin, D. Pegg, Phys. Rev. **175**, 44 (1968)
21. C. Darwin, Proc. Roy. Soc. Lond A **118**, 147 (1928)
22. C. Doran, A. Lasenby, *Geometric Algebra for Physicists* (Cambridge University Press, Cambridge, UK)
23. F.J. Dyson, Phys. Rev. **73**, 617 (1947)
24. J. Eichler, Phys. Rep **193**, 166 (1990)
25. J.B. French, V.F. Weisskopf, Phys. Rev. **75**, 1240 (1949)
26. M. Gel'Fand, R. Minlos, Z. Shapiro, *Representation of the rotations and Lorentz groups* (Pergamon Oxford, 1963)
27. W. Gordon, Z. Phys. **58**, 11 (1928)
28. S. Gull, in *The Electron*, ed. by D. Hestenes, A. Weingartshofer (Kluwer, Dorderecht, 1991), p. 233

132   References

29. R. Gurtler, Dissertation, Arizona State University, 1972
30. R. Gurtler, D. Hestenes, J. Math. Phys. **16**, 573 (1975)
31. F. Halbwachs, J.M. Souriau, J.P. Vigier, J. Phys. et le Radium **22**, 393 (1961)
32. H. Hall, Rev. Mod. Phys. Rev. **8**, 358 (1936)
33. S. Haroche, M. Brune, J. M. Raimond, in *Atomic Physics* 12, ed. by J.C. Zorn, R.R. Lewis (American Institute of Physics, New York, 1991)
34. D. Hestenes, *Space Time Algebra* (Gordon and Breach, New York, 1966)
35. D. Hestenes, J. Math. Phys. **8**, 798 (1967)
36. D. Hestenes, J. Math. Phys. **14**, 893 (1973)
37. D. Hestenes, Ann.J. Math. Phys. **71**, 718 (2003)
38. A. Ichihara, J. Eichler, Atomic Data and Nuclear Data Tables **74**, 1 (2000)
39. G. Jakobi, G. Lochak, C. R. Acad. Sci. (Paris) **243**, 234 (1956)
40. M. Kroll, W. Lamb, Phys. Rev. **75**, 388 (1949)
41. H. Krüger, Elektrodynamik, Universitat Kaiserslauter (1988)
42. W.E. Lamb Jr, R.C. Retherford, Phys. Rev. **72**, 241 (1947)
43. L.D. Landau, E.M. Lifshiftz, *Quantum Mechanics*, vol. 4 (Pergamon London, 1971)
44. A. Lasenby, C. Doran, S. Gull, in *Spinors, Twistors, Clifford Algebras and Quantum Deformations*, Z. Oziewicz, B. Jancewicz, A. Borowiec ed. by (Kluwer, dorderecht, 1993) p. 233
45. W. Magnus, F. Oberhettinger, R.P. Soni, *Formulas and Theorems for the Special Functions of Mathematical Physics* (Springer, Berlin, 1966)
46. H. Margenau, Phys. Rev. **57**, 383 (1940)
47. A. Messiah, *Mécanique Quantique*, Tome II (Dunod, Paris, 1969)
48. P. Quilichini, C. R. Acad. Sci. (Paris) B **273**, 829 (1971)
49. M.E. Rose, *Relativistic Electron Theory* (Wiley, New York, 1961)
50. L. Schiff, *Quantum Mechanics* (Mc Graw-Hill, New York, 1955)
51. J. Seke, Mod. Phys. Lett. B **7**, 1287 (1993)
52. J. Seke, Z. Phys. D **29**, 1 ((1994)
53. J. Seke, Physica A **233**, 469 (1996)
54. A. Sommerfeld, *Atombau und Spectrallinien* (Wieweg Brauhschweig, Berlin, 1960)
55. A. Sommerfeld, G. Schur Ann. Phys. Leipzig **4**, 409 (1930)
56. T. Takabayasi, Sup. Prog. Theor. Phys. **4**, 1 (1957)
57. J. Yvon, J. Phys. Radium **8**, 18 (1940)

Index

Springer Series on
ATOMIC, OPTICAL, AND PLASMA PHYSICS

Springer Series on
ATOMIC, OPTICAL, AND PLASMA PHYSICS